Diss. ETH No. 21528

Enabling Dependable Communication in Cyber-Physical Systems with a Wireless Bus

A dissertation submitted to
ETH Zurich

for the degree of
Doctor of Sciences

presented by
FEDERICO FERRARI
M.Sc. ALaRI, USI
born January 6, 1981
citizen of Italy

accepted on the recommendation of
Prof. Dr. Lothar Thiele, examiner
Prof. Dr. Prabal Dutta, co-examiner
Prof. Dr. Luca Mottola, co-examiner

2013

Institut für Technische Informatik und Kommunikationsnetze
Computer Engineering and Networks Laboratory

TIK-SCHRIFTENREIHE NR. 141

Federico Ferrari

Enabling Dependable Communication in Cyber-Physical Systems with a Wireless Bus

Eidgenössische Technische Hochschule Zürich
Swiss Federal Institute of Technology Zurich

A dissertation submitted to
ETH Zurich
for the degree of Doctor of Sciences

Diss. ETH No. 21528

Prof. Dr. Lothar Thiele, examiner
Prof. Dr. Prabal Dutta, co-examiner
Prof. Dr. Luca Mottola, co-examiner

Examination date: October 18, 2013

ISBN 978-1492251033
DOI 10.3929/ethz-a-010000257

To grandma and grandpa.
(I know you'd be proud!)

Abstract

Cyber-physical systems (CPSs) are physical and engineered systems believed to radically transform how we interact with the physical world. By tightly integrating computation, low-power wireless communication, and physical processes, these systems realize safety-critical control loops—with physical processes affecting computation and vice versa—in scenarios where traditional systems are hardly applicable. Potential CPS applications include healthcare, factory automation, and smart structures.

The safety-critical nature of most CPS applications demands highly dependable system operation. However, it is currently not possible to apply to cyber-physical systems established concepts for the design and validation of dependable distributed systems. These concepts require guarantees (e.g., on message delivery orderings) that existing CPS communication protocols do not provide.

It is indeed extremely challenging to guarantee message delivery in low-power wireless networks, due to, for example, severe computation and memory constraints of typical CPS embedded devices, multi-hop wireless communication, and the need of satisfying also requirements on energy efficiency. State-of-the-art solutions try to overcome these challenges by involving in the exchange of messages as few nodes as possible, but they typically operate only in a best-effort manner.

By contrast, we conjecture in this thesis that it is possible to enable dependable yet efficient communication in cyber-physical systems by employing a *wireless bus*—a time-triggered communication infrastructure where, similar to protocols for (wired) safety-critical embedded systems, nodes are time-synchronized and communicate as if they were connected by a shared bus. In particular, we implement three main building blocks contributing towards a dependable wireless bus:

- We design Glossy, a flooding architecture that provides fast and highly reliable one-to-all communication in multi-hop low-power wireless networks, while also accurately time-synchronizing all devices. Glossy exploits synchronous transmissions of the same packet, and does not require nodes to maintain any knowledge of the network topology. Experimental results from three testbeds show that Glossy delivers messages within a few milliseconds and with probabilities above 99.99 % in most scenarios, while also providing global time synchronization with sub-microsecond accuracy.

- We present the Low-Power Wireless Bus (LWB), a wireless bus that maps all traffic demands on Glossy floods and globally schedules every flood, thus effectively turning a multi-hop wireless network into an infrastructure similar to a shared bus where all nodes are potential receivers of all data. Therefore, LWB inherently supports one-to-many, many-to-one, and many-to-many traffic without keeping any topology-dependent state at the nodes. Results from four testbeds show that LWB performs comparably or significantly better than seven state-of-the-art many-to-one and many-to-many protocols, adapts efficiently to traffic loads and network topologies varying over time, ensures fair bandwidth allocation, and supports mobile nodes without performance loss.

- We finally verify the validity of our conjecture by developing VIRTUS, a wireless bus that extends LWB's best-effort operation and provides virtual synchrony guarantees. By implementing atomic multicast and view management, VIRTUS ensures that non-faulty nodes see the same events in the same order despite possible communication failures or node crashes. Virtually-synchronous executions allow to apply to cyber-physical systems established methods for fault tolerance based on replication techniques. Testbed results show that VIRTUS implements virtual synchrony at a marginal cost compared with LWB, and is significantly more energy-efficient than existing best-effort multicast protocols for low-power wireless networks.

Sommario

Si ritiene che i sistemi ciber-fisici (CPSs) possano cambiare radicalmente il modo in cui interagiamo con il mondo fisico intorno a noi. Integrando in modo stretto computazione, comunicazione senza fili a bassa potenza e processi fisici, questi sistemi realizzano circuiti di controllo critici per la sicurezza, dove processi fisici e computazione si influenzano a vicenda, e in situazioni in cui sistemi più tradizionali sono difficilmente applicabili. Possibili applicazioni CPSs includono assistenza sanitaria, automazione industriale e strutture intelligenti.

Poiché la maggior parte delle applicazioni CPSs sono critiche per la sicurezza, è fondamentale che questi sistemi funzionino in modo affidabile. Tuttavia, non è al momento possibile applicare ai sistemi ciber-fisici concetti consolidati per la progettazione e la validazione di sistemi distribuiti affidabili. Questi concetti richiedono garanzie (per esempio sulla consegna ordinata dei messaggi) che i protocolli di comunicazione esistenti non forniscono.

È infatti estremamente arduo garantire la consegna di messaggi in reti senza fili a bassa potenza, per esempio a causa della limitata capacità di computazione e di memoria dei sistemi *embedded* (incorporati) tipici dei CPSs, dell'uso di comunicazione senza fili di tipo *multi hop* (multi salto) e della necessità di soddisfare anche requisiti di efficienza energetica. Le soluzioni attuali cercano di superare questi ostacoli coinvolgendo nello scambio di messaggi il minor numero possibile di nodi, ma solitamente operano solo in modalità *best effort* (senza garanzie).

In questa tesi ipotizziamo che è invece possibile ottenere una comunicazione affidabile oltre che efficiente nei sistemi ciber-fisici utilizzando un *wireless bus* (bus senza fili)—un'infrastruttura *time triggered* (attivata dall'avanzare del tempo) dove, similmente a protocolli per sistemi embedded cablati critici per la sicurezza, i nodi sono sincronizzati e comunicano come se fossero collegati da un unico bus condiviso. In particolare, implementiamo tre mattoni principali che concorrono a realizzare un wireless bus affidabile:

- Ideiamo Glossy, un'architettura *flooding* (per l'inondazione di pacchetti) che consente ad un nodo di distribuire pacchetti in modo veloce ed altamente affidabile a tutti gli altri nodi di una rete senza fili a bassa potenza e multi hop, oltre a sincronizzare accuratamente tutti i dispositivi. Glossy sfrutta trasmissioni contemporanee dello stesso pacchetto e non richiede che i nodi abbiano alcuna conoscenza della topologia della rete. Esperimenti su tre *testbeds* (banchi di prova per reti senza fili) mostrano che nella maggior parte dei casi Glossy consegna messaggi entro pochi millisecondi e con probabilità maggiori del 99.99 %, fornendo anche una sincronizzazione globale con una precisione al di sotto del microsecondo.

- Presentiamo il Low-Power Wireless Bus (LWB), un wireless bus che usa Glossy per tutte le comunicazioni e che pianifica in modo globale ogni scambio di messaggi, trasformando di fatto una rete di comunicazione senza fili e multi hop in un'infrastruttura simile ad un bus condiviso. LWB perciò supporta intrinsecamente traffico di tipo uno-verso-tutti, tutti-verso-uno e tutti-verso-tutti senza dover mantenere nessun tipo di stato che dipenda dalla topologia della rete. Esperimenti su quattro testbeds mostrano che LWB ha prestazioni simili o molto migliori di sette protocolli esistenti di tipo tutti-verso-uno e tutti-verso-tutti, si adatta in modo efficiente a carichi di traffico e topologie di rete variabili, assicura un'allocazione equa della banda e supporta nodi mobili senza perdita di prestazioni.

- Verifichiamo la validità della nostra ipotesi sviluppando Virtus, un wireless bus che estende la modalità best effort di LWB e fornisce garanzie di *virtual synchrony* (sincronia virtuale). Implementando *atomic multicast* e *view management*, Virtus assicura che i nodi non guasti vedano gli stessi eventi nello stesso ordine, nonostante possibili mancate ricezioni o crash di nodi. Esecuzioni di tipo virtual synchrony permettono di applicare ai sistemi ciber-fisici metodi consolidati per la tolleranza ai guasti basati su tecniche di replicazione. Esperimenti su testbeds mostrano che Virtus implementa virtual synchrony con un costo marginale rispetto a LWB ed è sensibilmente più efficiente di protocolli multicast e best effort per reti senza fili a bassa potenza.

Acknowledgments

First of all, I would like to thank my advisor Prof. Dr. Lothar Thiele for giving me the opportunity to work on this thesis. Thank you for your valuable support, your patience throughout these years, and your inspiring advices. I would also like to thank Prof. Dr. Prabal Dutta and Prof. Dr. Luca Mottola for co-examining this thesis.

This work would have not been possible without the continuous, fruitful discussions with my colleague Marco Zimmerling, and his significant contributions to the work presented in this thesis. I would also like to express my gratitude to Dr. Jan Beutel, Prof. Dr. Olaf Landsiedel, Prof. Dr. Luca Mottola, and Dr. Olga Saukh for the successful collaborations during these years.

I am grateful to have been part of a motivating and friendly team at the Computer Engineering Laboratory. I have met many new friends here, which I would like to thank for making the time during my PhD special and for standing my sometimes weird jokes.

Finally, I would like to thank my parents, my uncle, my sister, and my "bro" for their constant support. Most importantly, thank you Monica for supporting and standing by me throughout these years, and for encouraging me especially during the difficult times!

The work presented in this thesis was funded in part by Nano-Tera and the National Competence Center in Research on Mobile Information and Communication Systems (NCCR-MICS), a center supported by the Swiss National Science Foundation under grant number 5005-67322.

Contents

List of Figures

List of Tables

1

Introduction

Cyber-physical systems (CPSs) are physical and engineered systems whose operation is controlled by a computing and communication core. In these systems, embedded computers and networks monitor and control physical processes, usually with feedback loops where physical phenomena affect computation and vice versa. This tight integration of computation, communication, and physical processes allows cyber-physical systems to realize safety-critical control loops in scenarios where traditional systems are hardly applicable.

Examples of CPS applications include high-confidence medical devices and systems, assisted living, traffic control and safety, advanced automotive systems, factory automation, energy conservation, environmental control, critical infrastructure control (e.g., electric power, water resources, and communications systems), distributed robotics (e.g., telepresence and telemedicine), defense systems, manufacturing, and smart structures [Lee08, RLSS10]. It is commonly believed that these systems are going to radically transform how we interact with the physical world around us, similar to how the internet transformed the way humans interact with one another [Lee08, RLSS10, Sch12, SLMR05].

The safety-critical nature of most cyber-physical systems, however, raises several concerns [SLMR05]. Deployed systems must guarantee highly dependable operation against unpredictable real-world dynamics. The most important CPS requirements are concisely summarized by Rajkumar et al. as follows [RLSS10]:

> Cyber-physical systems must operate dependably, safely, securely, efficiently, and in real-time.

Careful designs providing guarantees on the system behavior are thus required. However, applying established concepts for the design and validation of dependable distributed systems to cyber-physical systems is currently not possible, as these concepts require guarantees that existing CPS communication protocols do not provide. Such guarantees notably include, for example, well-defined message delivery orderings that facilitate the implementation of replicated functionality, as well as failure handling mechanisms [Sch90, KDK⁺89]. The fact that communication among embedded devices is one of the major challenges in the design of cyber-physical systems was also pointed out by Lee [Lee08]:

> Most safety-critical embedded systems are closed "boxes" that do not expose the computing capability to the outside. The radical transformation that we envision comes from networking these devices. Such networking poses considerable technical challenges.

Nevertheless, we conjecture in this thesis that it is possible to enable dependable communication in cyber-physical systems by employing approaches similar to those currently used in the design safety-critical embedded systems (e.g., in automotive and avionics). We first illustrate in Section 1.1 the main challenges we must confront in order to enable dependable communication. After summarizing in Section 1.2 current state of the art of CPS protocols designed for dependability, we then formalize our conjecture in Section 1.3, along with an overview of how we verify its validity throughout the remaining chapters.

1.1   Challenges in Low-Power Wireless

At the core of typical cyber-physical systems are embedded devices with minimal computation and wireless communication capabilities. By gathering data from the environment through (integrated or external) *sensors* and taking actions on it through (integrated or external) *actuators*, these resource-constrained devices effectively realize the interactions with the physical processes required by most CPS applications [Sta08]. Nevertheless, the limited amount of resources available on these devices poses several important challenges to the design of wireless communication protocols, ultimately hampering the possibility to provide guarantees on their operation. We now highlight the three most important sources of challenges.

Resource-constrained embedded devices. Typical platforms for wireless embedded devices feature a low-power microcontroller (MCU) and a short-range, low-rate wireless radio within an area of a few square

Figure 1.1:   A TelosB (aka Tmote Sky) embedded device features a microcontroller and a low-power wireless radio within a few square centimeters.

centimeters.   For example, a TelosB device [PSC05] (also known as Tmote Sky) like the one shown in Figure 1.1 employs a 16-bit MSP430 microcontroller operating at frequencies up to 8 MHz [Texe] and a CC2420 wireless radio compliant with the IEEE 802.15.4 standard [IEE03] and transmitting on the 2.4 GHz band at a fixed rate of 250 kbps [Texb].

The storage capabilities of these devices are also extremely limited. For instance, the MCU of a TelosB device features only 10 kB of RAM and 48 kB of program memory; an external flash chip provides 1 MB of non-volatile memory. Recent advances in the field of sensing platforms have shown that it is even feasible to integrate computation, communication, storage, and sensing within a single chip of only one cubic-millimeter [LBL+13]. While this means that the smart dust vision [KKP99] may soon become reality and thus open opportunities for a plethora of new CPS applications, it entails that these embedded devices are likely to remain extremely resource-constrained also in the near future. Such scarcity of resources, especially in terms of storage, severely limits the possibility for communication protocols to perform basic operations on the device, such as buffering multiple messages.

Multi-hop wireless communication. The low-power wireless radios available on these platforms have ranges limited to up to a few tens of meters indoors (up to a few hundreds of meters outdoors), yet several CPS applications require to cover significantly larger areas (e.g., factory automation and smart structures). For this reason, designers usually employ *multi-hop* wireless networks, where each device—a *node* of such networks—can directly exchange data only with a subset of other nodes: those that lie within its communication range.

Figure 1.2 shows an example of a typical CPS multi-hop *low-power wireless network* consisting of several sensor and actuator nodes. In such

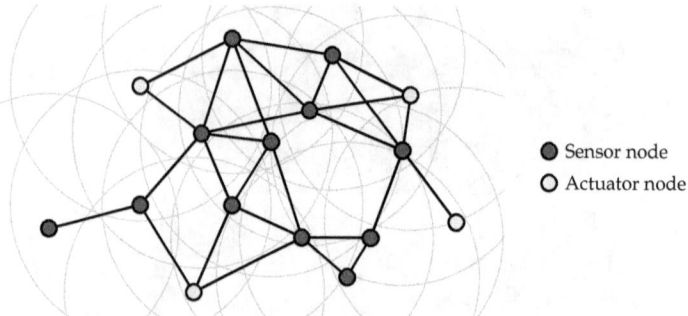

Figure 1.2:   Example of a multi-hop low-power wireless network with twelve sensor nodes and four actuator nodes. Lines represent physical communication links between pairs of nodes; circles represent nodes' communication ranges.

networks, nodes relay messages on each other's behalf, thus enabling communication between nodes outside of each other's communication range. However, the quality of physical *communication links* between pairs of nodes fluctuates during operation, for example, due to multipath effects, obstacles, and external interference [SDTL10]. As a result, the network topology changes continuously—even in the absence of mobility—making it challenging for communication protocols to reliably, efficiently, and timely deliver messages across multiple hops [AY05].

Energy efficiency is paramount. Many CPSs are designed to operate continuously—possibly unattended—for periods that may range from a few weeks to several years [BGH+09, CMP+09, CCD+11], yet the amount of energy storage available for these embedded devices is often extremely limited due to size, portability, cost, and physical reasons. A fully-active TelosB device, for example, would drain two common AA batteries within a few days. Because the wireless radio is the component that typically draws most of the current, communication protocols must ensure that the radio operates in a low-power sleep mode for a large fraction of time, in order to meet given requirements on system lifetime. This significantly complicates protocol design, as any pair of nodes can communicate only when both devices have their wireless radios turned on.

1.2   Existing Communication Support

A large body of work on low-power wireless communication protocols is available in the literature [AY05, Lan08], as plenty of different strategies can be employed to tackle the challenges above. Some of these protocols have been successfully used in past or ongoing wireless sensor network deployments, usually meeting given lifetime and reliability requirements [BGH+09, CMP+09, CCD+11]. Nevertheless, their best-effort operation prevents providing guarantees on system dependability. We exemplify this issue based on an existing CPS deployment.

A concrete CPS example. The TRITon project uses closed-loop control for adaptive lighting in road tunnels to improve their safety. Battery-powered TelosB-like sensor nodes report periodic light readings to a central controller running on embedded hardware, which closes closes the loop by setting the lamp intensity to match a legislated curve [CCD+11]. Using current communication protocols, however, TRITon designers cannot provide dependability assurances. Moreover, the centralized controller represents a single-point of failure. Designers wish to address these concerns, for example, by replicating the control logic across devices, as done in typical fault-tolerant distributed systems [Sch90].

By employing replication, the TRITon network would look similar to the one in Figure 1.2, where the replicated controllers would correspond to the actuator nodes. Well-established approaches for fault tolerance would require these controllers to process the same messages from the sensor nodes in the same order [Sch90]. However, no existing low-power wireless protocol for many-to-many communication guarantees ordered delivery. It is indeed extremely difficult to achieve such type of global coordination in these networks, because nodes: *i)* can only buffer a very limited number of messages due to the memory shortage, *ii)* need to rely on links with fluctuating quality and thus on a time-varying set of intermediate devices, and *iii)* must reduce communication to a minimum in order to save energy and meet the requirements on system lifetime.

Guarantees only in specific scenarios. Most existing communication protocols indeed operate in a best-effort manner—their design is optimized towards non-functional properties, such as energy consumption [AY05]. In order to involve a minimal number of nodes and thus save energy, state-of-the-art solutions typically split end-to-end interactions into multiple independent single-hop transmissions among subsets of nodes [AY05, BvRW07, GFJ+09]. The result is that these protocols are designed to support only specific scenarios, such as a many-to-one traffic pattern in static networks. Most importantly, the time-varying nature of low-power wireless network topologies makes it extremely difficult for these protocols to ensure that messages are delivered reliably, timely, or in order.

Some solutions exist to provide guarantees in specific scenarios. Structural health monitoring applications [CMP+09], for example, often require guaranteed message delivery from multiple sensor nodes to a single data sink. Protocols like RCRT [PG07] and several ad-hoc solutions [CMP+09] provide such functionality. Real-time scheduling algorithms [SXLC10] allow to meet end-to-end deadlines when using WirelessHART [SHM⁺08], an open standard for industrial process monitoring and control [SHM⁺08]. Nevertheless, these protocols support only many-to-one traffic patterns. Common replication techniques for fault tolerance and the sense-process-actuate cycles of typical CPS applications, however, require many-to-many interactions [PSLN⁺12, Sch90]. Unfortunately, existing low-power multicast protocols typically provide only best-effort operation [AY05, MP11].

1.3 Taking a Different Stand with a Wireless Bus

In this thesis, we take a radically different approach to achieve global coordination and ultimately enable dependable communication in cyber-physical systems. Inspired by existing bus-based solutions for the design of (wired) safety-critical embedded systems [Rus01, Kop11, KG93, MT06], we formulate the following conjecture:

Conjecture. *We can enable dependable yet efficient communication in cyber-physical systems by employing a* wireless bus—*a time-triggered communication infrastructure for multi-hop low-power wireless networks where nodes are time-synchronized and communicate as if they were connected by a shared bus.*

A wireless bus is in stark contrast to state-of-the-art solutions for low-power wireless networks, as the latter typically aim to involve in the communication as few nodes as possible in order to achieve high energy efficiency. We instead believe that, by employing time-triggered executions where the entire operation is driven by a common notion of time shared across all nodes, a wireless bus would enable dependable communication while satisfying also typical requirements for energy efficiency. Nevertheless, we note that our conjecture is in line with similar thoughts expressed by Lee in [Lee08, Lee09]:

> What aspect of networking technologies such as CAN busses in manufacturing systems and FlexRay in automotive applications should or could be important in larger-scale networks? To be specific, recent advances in time synchronization across networks promise networked platforms that share a common notion of time to a known precision. How would that change how distributed cyber-physical applications are developed?

In this thesis, we implement three main building blocks contributing towards a dependable wireless bus. Figure 1.3 illustrates how these building blocks turn a typical CPS low-power wireless network like the one in Figure 1.2 into a wireless bus with delivery guarantees.

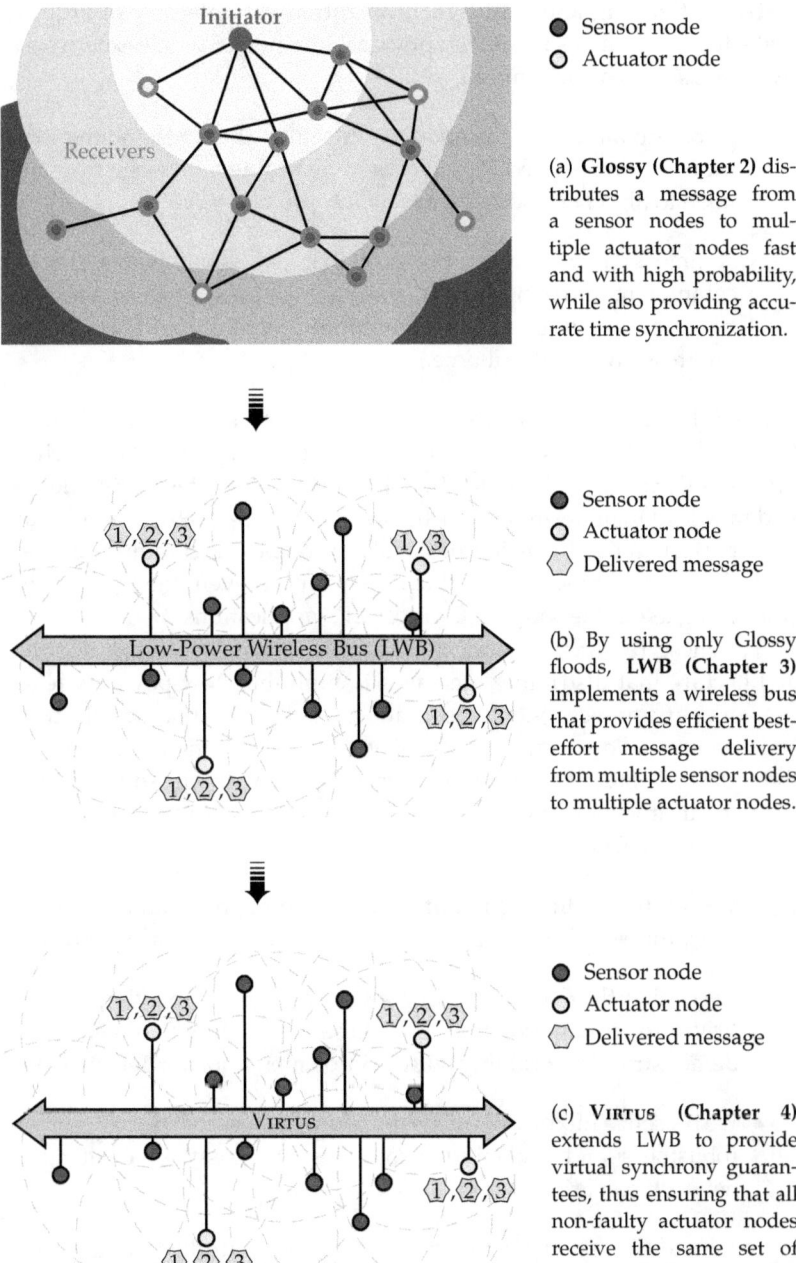

Figure 1.3: The building blocks presented in this thesis turn the low-power wireless network of Figure 1.2 into a wireless bus with delivery guarantees.

Network flooding and time synchronization (Chapter 2). In order to make the wireless bus feasible in practical low-power wireless networks, two basic services are required:

- A one-to-all communication architecture that, resembling data transfers on a shared bus, allows to distribute a message from one node to all other nodes *fast and with high probability*.

- A time synchronization protocol that makes all nodes share a common notion of time *with high accuracy*, necessary to enable time-triggered executions where the entire operation is driven by the progression of a global time.

State-of-the-art solutions provide these services separately. A wireless bus would thus require to run two protocols in parallel, such as Trickle [LPCS04] and FTSP [MKSL04], which complicates system design and may lead to undesired protocol interactions [CKJL09]. Moreover, the poor performance of existing protocols for one-to-all communication—typically referred to as *network flooding*—would prevent the wireless bus from being a feasible solution for most CPS applications.

For these reasons, we introduce Glossy, a novel network flooding architecture that distributes messages in multi-hop low-power wireless networks at unprecedented speed and reliability, and accurately time-synchronizes all devices. Glossy achieves this by making nodes transmit the same IEEE 802.15.4 packet *synchronously*, and without requiring them to maintain any knowledge of the network topology. We detail the Glossy mechanisms in Chapter 2, where:

- We study the timing requirements to make baseband signals of synchronous IEEE 802.15.4 transmissions interfere constructively.

- We design the Glossy flooding architecture by employing a radio-driven execution model in order to satisfy such requirements, and demonstrate its feasibility with an implementation for TelosB nodes.

- We present a mixture of stochastic and worst-case models to analyze robustness and scalability properties of Glossy in challenging network scenarios.

We evaluate Glossy using experiments under controlled settings and on three wireless sensor testbeds. Results show that in most scenarios nodes receive a flooding message within a few milliseconds and with a probability higher than 99.99 %. Moreover, Glossy time-synchronizes an entire network with sub-microsecond accuracy.

A concrete wireless bus (Chapter 3). We leverage Glossy to design a concrete wireless bus protocol, the Low-Power Wireless Bus (LWB). By employing only Glossy floods for communication, LWB effectively turns a multi-hop wireless network into an infrastructure similar to a shared bus where all nodes are potential receivers of all messages. As a result, LWB natively supports multiple traffic patterns, such as one-to-many, many-to-one, and many-to-many. Moreover, nodes exchange messages under the illusion of being in each other's communication range, as Glossy hides from them the complexity of the actual multi-hop network topology.

LWB's time-triggered operation dictates that nodes access the bus (i.e., transmit a message with Glossy) according to a global communication schedule. A dedicated host node computes this schedule online based on current traffic demands and periodically distributes it to all nodes. As described in Chapter 3, we complement LWB with mechanisms that:

- Ensure a fair allocation of bandwidth across all nodes and support traffic demands that change at runtime.

- Support nodes dynamically joining and leaving the system (e.g., due to node failures or disconnections).

- Resume communication after a host failure, thus overcoming single point of failure problems.

We implement a LWB prototype and demonstrate the effectiveness of these mechanisms on real networks. Moreover, we use four testbeds to compare the same LWB prototype with seven state-of-the-art protocols. Results from 838 hours of tests show that:

- LWB efficiently supports multiple traffic patterns. For example, it outperforms existing multicast solutions for many-to-many communication, at times by orders of magnitude. Moreover, LWB performs comparably or significantly better than state-of-the-art many-to-one protocols.

- LWB is resilient to topology changes and supports mobility with no performance loss. Because Glossy maintains no topology-dependent state at the nodes, no state reconfigurations are required when the topology changes, for example, due to external interference, node failures, or mobility. For instance, LWB delivers more than 99 % of messages at very low energy costs also when one or multiple nodes are roaming.

- Thanks to its centralized and time-triggered operation, LWB is energy-efficient also in networks with hundreds of nodes, despite it forces all nodes to participate in every message exchange.

A wireless bus with virtual synchrony guarantees (Chapter 4). Nodes in LWB receive messages with probabilities close to 100 % in most scenarios. Nevertheless, LWB is a best-effort protocol that cannot ensure successful message delivery. According to our conjecture in page 6, however, its bus-like operation should enable global coordination among nodes, thus rendering it possible to design protocols with delivery guarantees.

To verify the validity of our conjecture, we design VIRTUS, a communication protocol that builds on LWB and provides virtual synchrony guarantees. Virtually-synchronous executions are typically exploited by fault tolerance methods based on replication techniques [Sch90], as it is essentially ensured that every non-faulty replica sees the same events in the same order [BJ87].

As we detail in Chapter 4, VIRTUS provides virtual synchrony guarantees by extending LWB with two main services:

- A view management service, which keeps nodes informed of the current group of non-faulty nodes, while also managing changes in the group, for example, in response to failures.

- An atomic multicast service, which delivers messages to non-faulty nodes reliably and with total order.

We formally prove that our design does enable virtually-synchronous executions—in VIRTUS, any non-faulty node sees the same events (i.e., nodes joining and leaving the group and message deliveries) in the same order regardless of possible communication failures or node crashes. We also complement virtual synchrony with further policies that ensure FIFO ordered delivery.

We evaluate a prototype of VIRTUS on two testbeds, and show that it provides virtual synchrony at a marginal cost compared with LWB's best-effort operation. As a matter of fact, VIRTUS guarantees ordered delivery while being significantly more energy-efficient than existing multicast protocols for low-power wireless networks. To the best of our knowledge, VIRTUS is the first protocol to provide formally-proven virtual synchrony atop similarly resource-constrained hardware.

2

Glossy:
Efficient Network Flooding
and Time Synchronization

Network flooding and time synchronization are two fundamental services in multi-hop low-power wireless networks, as they form the basis for a wide range of applications and network operations. More specifically, in this thesis we require both services to render it possible to design a wireless bus that enables dependable communication in cyber-physical systems. Nevertheless, many existing data collection applications based on wireless sensor networks (WSNs) also rely on a seamless coexistence of these services. For instance, high-rate data collection systems synchronize nodes to correlate measurements, and use flooding to adjust sampling rates and trigger data downloads [WALJ+06, CMP+09]. Most of these applications run two protocols in parallel (e.g., Trickle [LPCS04] and FTSP [MKSL04]), which complicates their design and may cause protocol interactions that impair system performance [CKJL09]. A flooding service that implicitly synchronizes all nodes in the network could effectively avoid these problems.

Such an integrated service should flood packets as fast as possible to reduce inaccuracies introduced by clock drift [LSW09]. Moreover, rapid flooding can enhance the performance of several applications [LW09]. In surveillance systems, for example, a node detecting an event needs to quickly wake up all other nodes to initiate group formation and collaborative signal processing [LWHS02].

Challenges.   Rapid flooding is difficult in multi-hop low-power wireless networks where packet loss is a common phenomenon [ZG03]. Retransmissions to recover lost packets help overcome this problem. However, simple broadcasting results in serious medium contention, known as the broadcast storm problem [NTCS99].   To reduce the transmission overhead, nodes need to acknowledge broadcasts using sophisticated modulation schemes [DSGS09], encode redundant information into packets prior to transmission [RZS+08], or collect substantial information from neighboring nodes to decide whether a retransmission is necessary [WC02]. Therefore, loss recovery generally sacrifices latency and energy for an increased reliability.

Alternatively, reliability can be improved by reducing the risk of packet loss in the first place.   One possible approach is to schedule broadcasts so that they do not interfere with each other.   However, determining an interference-free broadcast schedule is an NP-complete problem [ET90] and subject to sudden topology changes.

In fact, due to the capture effect, a node can receive a packet despite interference from other wireless transmitters [LF76].   While the capture effect helps improve flooding efficiency, it suffers from scalability problems in areas of high node density: the probability of receiving a packet decreases considerably as the number of synchronous transmissions increases [LW09].

Contribution and road-map.   To tackle the issues above, this chapter proposes Glossy, a new flooding architecture for low-power wireless networks.   Glossy considers interference an advantage rather than a problem.   Unlike previous work, it makes *the baseband signals of synchronous transmissions of the same packet interfere constructively*, allowing receivers to decode the packet even in the absence of capture effects. In this way, Glossy achieves a flooding reliability above 99.99 % and approaches the theoretical lower latency bound across diverse node densities and network diameters. Moreover, Glossy provides network-wide time synchronization for free, since it implicitly synchronizes nodes as the flooding packet propagates through the network.

This chapter makes the following contributions:

- We study in Section 2.1 why and under which conditions the baseband signals of synchronous transmissions of the same packet interfere constructively. Our analysis reveals a strong dependence on the modulation scheme. Based on this insight, we show that the temporal offset among synchronous IEEE 802.15.4 transmitters must not exceed $0.5\,\mu s$ to make the baseband signals interfere constructively with high probability.

- We introduce Glossy, a new flooding architecture for low-power wireless networks. Glossy exploits synchronous transmissions, time-synchronizes nodes, and decouples flooding from other network activities. We give an overview of Glossy's design in Section 2.2, and detail its radio-driven execution model in Section 2.3.

- We demonstrate in Section 2.4 the feasibility of Glossy with an implementation in Contiki [Conb, DGV04] based on TelosB sensor nodes [PSC05]. We describe how our implementation reduces time uncertainties on the nodes during packet relaying, and give guidelines for porting Glossy to other popular hardware platforms.

- We present in Section 2.5 a mixture of stochastic and worst-case models to analyze the robustness of our techniques in generating constructive interference. Applying these models to our implementation, we find that Glossy satisfies the $0.5\,\mu s$ requirement with a probability higher than 99.9 % for 30 synchronous transmitters.

In Section 2.6, we evaluate Glossy using experiments under controlled settings and on three wireless sensor testbeds: MOTELAB [WASW05], TWIST [HKWW06], and DSN [DBK$^+$07]. For example, we find that Glossy achieves an average time synchronization error below $0.4\,\mu s$, even at nodes that are eight hops away from the initiator of a flood. On TWIST, at the lowest transmit power that keeps the network fully connected, we observe that nodes receive an 8-byte flooding packet within 3 ms; nodes receive the packet with a probability above 99.99 %, and have their radios turned on for less than 10 ms during a flood.

In light of our contributions, Section 2.7 surveys related work, and Section 2.8 provides brief concluding remarks.

2.1 Synchronous Transmissions

Glossy exploits synchronous transmissions for efficient flooding in sensor networks. In this section, we investigate the conditions for making the baseband signals of synchronous transmissions of the same packet interfere in a constructive way, so that a receiver detects the packet with high probability. We give some background on wireless interference and the IEEE 802.15.4 modulation scheme, based on which we determine a timing requirement for generating constructive interference.

2.1.1   Background

The broadcast nature of wireless communications causes interference whenever spatially close stations transmit concurrently; that is, when they generate signals that overlap in time and space, and share the same frequency. Interference generally reduces the probability that a receiver correctly detects the information embedded into the signals, even when the signals carry the same information. In the following discussion we focus on *baseband* signals, that is, sequences of IEEE 802.15.4 symbols. As we show in Section 2.6, the superposition of several, possibly out-of-phase *carrier* signals allows for correct detection with high probability, especially when more than three nodes transmit synchronously.

Constructive and destructive interference. We say that interference is *constructive* if a receiver correctly detects the superposition of the baseband signals generated by multiple transmitters. By contrast, interference is *destructive* if it prevents a receiver from correctly detecting the superimposed baseband signals. Constructive interference has not been extensively exploited in low-power wireless networks, due to the difficulty of achieving sufficiently accurate synchronization and highly predictable software delays [SZHT07]. Several protocols [LW09] exploit instead the *capture effect*, which occurs when a wireless radio detects a frequency-modulated signal from one transmitter despite the interference from other transmitters. A radio may capture one signal when it is stronger than the others (*power capture* [LF76]), or when it starts being received earlier than the others (*delay capture* [DG80]). However, capture effects suffer from scalability problems when many transmissions overlap, leading to significant packet loss in dense networks [LW09].

Requirements for generating constructive interference strongly depend on the communication scheme, and especially on the modulation and the bit rate. We first review the specifications of the IEEE 802.15.4 standard. Then, we derive an *upper bound on the temporal displacement* Δ among multiple synchronous transmissions of the same packet that allows to correctly receive the packet with high probability due to constructive interference of the baseband signals.

IEEE 802.15.4 modulation. The IEEE 802.15.4 standard [IEE03] for wireless devices operating in the 2,450 MHz band employs an offset quadrature phase-shift keying (O-QPSK) modulation scheme with half-sine pulse shaping, which is equivalent to minimum-shift keying (MSK). Binary data are converted into a modulated analog signal using a three-step conversion process, as shown in Figure 2.1.

First, data are grouped into 4-bit *symbols*. Each symbol is then mapped into a *pseudo-random noise (PN) sequence* of 32 bits, where each bit of such a

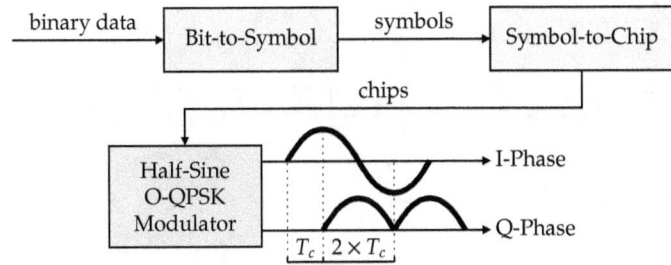

Figure 2.1: IEEE 802.15.4 modulation. Using a three-step process, binary data are converted into a modulated signal.

sequence is called *chip*. PN sequences add redundancy, and relate to each other through cyclic shifts and conjugation of chips. In a last step, each PN sequence is modulated onto the carrier signal using O-QPSK with half-sine pulse shaping. That is, even-indexed chips are modulated onto the *in-phase (I)* carrier, odd-indexed chips onto the *quadrature-phase (Q)* carrier. Q-phase chips are delayed by $T_c = 0.5\,\mu s$ with respect to I-phase chips to get a $\pi/2$ phase change. Overall, a new chip is transmitted every T_c, leading to a transmission rate of 250 kbps.

Demodulation at a receiver follows the opposite flow: each half-sine pulse is converted into a chip, and the resulting PN sequence into a symbol. Radios make only soft decisions for each chip [Texb]: the received PN sequence may contain non-binary values between 0 and 1. Hard decisions are made by selecting out of the 16 PN sequences the one that has the highest correlation with respect to the received PN sequence, and thus the corresponding symbol. In this step, the redundancy contained in a PN sequence increases the chances of a correct symbol detection, even in situations where some chips are not correctly received.

Next, we show how the IEEE 802.15.4 standard translates into a maximum temporal displacement Δ_{max} among multiple transmissions to generate constructive interference with high probability.

2.1.2 Generating Baseband Constructive Interference

Several studies [Chu87, YI85] estimate the bit error rate (BER) when receiving delayed replica of the same MSK signal (e.g., due to multipath effects). They show that the BER increases exponentially with the temporal displacement Δ among overlapping signals. However, as explained above, sensor network radios make hard decisions at the symbol level (i.e., on PN sequences of 32 consecutive chips). The redundancy included in each PN sequence helps tolerate decoding errors

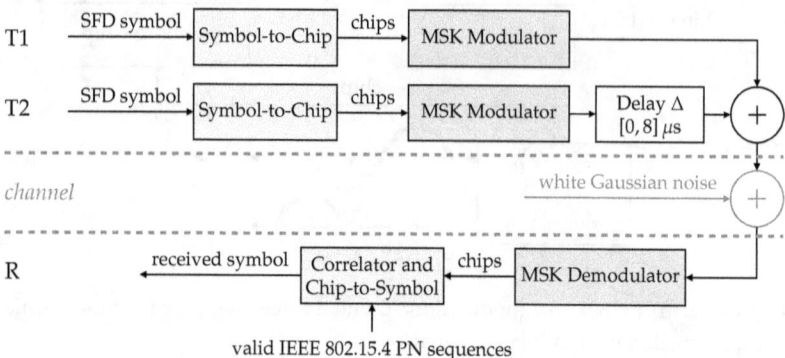

Figure 2.2: Setting of Matlab simulations. A receiver decodes the superposition of two IEEE 802.15.4-compliant SFD symbols, whose temporal displacement Δ varies between 0 μs and 8 μs with a granularity of 250 ns.

of single chips. Therefore, computing the error on a sequence of symbols provides a better estimation of the reception behavior of a sensor node.

We perform Matlab simulations to evaluate the maximum temporal displacement between two IEEE 802.15.4-compliant signals such that they interfere constructively with high probability. The setting of these simulations is shown in Figure 2.2. A receiver R decodes the superposition of two signals from transmitters T1 and T2. Each signal is generated by converting the start of frame delimiter (SFD) symbols specified by the IEEE 802.15.4 standard, first into a PN sequence and then into an MSK-modulated baseband signal. Both signals have the same amplitude, but the one from transmitter T2 is delayed by a variable displacement with 250 ns granularity in the interval [0, 8] μs. White Gaussian noise is added to the superimposed signal, resulting in a signal-to-noise ratio of -10 dB.

The receiver demodulates the superimposed signal. It then correlates each PN sequence with all 16 possible PN sequences specified by the IEEE 802.15.4 standard and chooses the one with the highest correlation. This procedure resembles the operations of a IEEE 802.15.4-compliant radio during a packet reception. Only if both SFD symbols are correctly decoded, the superimposed signal is considered correctly detected.

Figure 2.3 shows the fraction of correctly detected signals depending on the temporal displacement Δ, averaged over 1,000 experiments with different seeds for the random noise. The signal is always correctly detected when Δ = 0, which indicates that the noise is sufficiently low to allow for a correct detection. Even for Δ = 0.25 μs the signal is correctly detected in more than 98 % of the cases. However, the fraction of correct detections starts to decrease significantly for a displacement larger

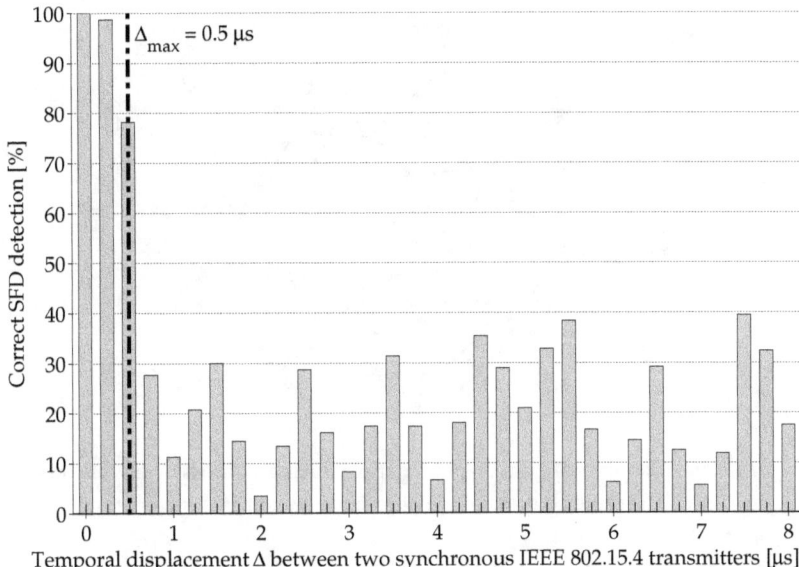

Figure 2.3: Synchronous IEEE 802.15.4 transmitters interfere constructively if the temporal displacement is smaller than $\Delta_{max} = 0.5\,\mu s$.

than $0.5\,\mu s$, which corresponds to the chip period T_c. Interestingly, for increasing Δ, the fraction experiences local minima when Δ is a multiple of $2 \times T_c$, that is, when different chips perfectly overlap. Between two local minima, the redundancy added by the PN sequences increases the chances for a correct signal detection, despite errors on single chips. We verify using various symbol sequences and noise seeds that these results are independent of the specific symbol sequence of the SFD.

These simulations show that the probability of a correct detection is very high when identical IEEE 802.15.4 signals are generated with a time displacement below $\Delta_{max} = 0.5\,\mu s$. A correct detection is entirely due to the modulation scheme and the redundancy encoded in PN sequences. On real nodes, capture effects can further increase the chances to correctly detect a packet, especially with high temporal or strength differences between the signals.

We show in the following how the design and the implementation of Glossy strive to satisfy the requirement of $\Delta_{max} = 0.5\,\mu s$, allowing nodes to receive packets even in the absence of beneficial capture effects (e.g., when many nodes transmit synchronously). Experimental results from three wireless sensor testbeds, described in Section 2.6, demonstrate that Glossy indeed achieves this goal, providing efficient network flooding across diverse node densities and network diameters.

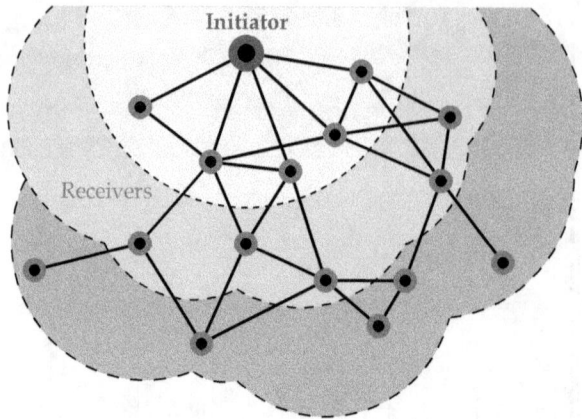

Figure 2.4: Example of packet propagation during a Glossy network flood. Nodes within the same color area relay the same packet at the same time.

2.2 Glossy Overview

This chapter introduces Glossy, a new flooding architecture for low-power wireless networks. Glossy incorporates three main techniques.

Synchronous transmissions. Wireless is a broadcast medium, creating the opportunity for nodes to overhear packets from neighboring nodes. Using Glossy, nodes turn on their radios, listen for communications over the wireless medium, and relay overheard packets immediately after receiving them. Since the neighbors of a sender receive a packet at the same time, they also start to relay the packet at the same time. This again triggers other nodes to receive and relay the packet. In this way, Glossy benefits from synchronous transmissions by quickly propagating a packet from a sender node (*initiator*) to all other nodes (*receivers*) in the network, as Figure 2.4 intuitively shows.

An important property of Glossy is that, besides the first transmission of the initiator, *the flooding process is entirely driven by radio events.* For instance, a node triggers a transmission only when the radio signals the completion of a packet reception. As explained in Section 2.1, synchronous transmissions must be properly aligned to enable a receiver to successfully decode the packet. Glossy's radio-driven execution is a key factor to meet this requirement.

Time synchronization. Glossy exploits the above flooding mechanism to implicitly time-synchronize the nodes. It embeds into each packet a 1-byte field, the *relay counter c*. The initiator sets $c = 0$ before the first transmission. Nodes increment c by 1 before relaying a packet.

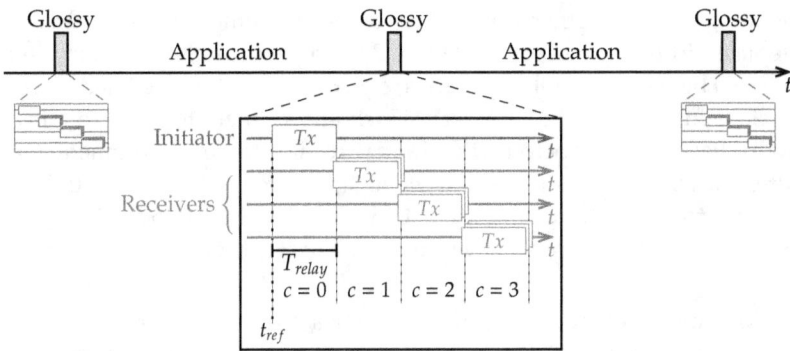

Figure 2.5: Glossy decouples flooding from other application tasks executing on the nodes. The reference time t_{ref} is the time at which the initiator starts a flood by transmitting a packet with relay counter $c = 0$. Nodes increment c by 1 before relaying a packet, thus they transmit packets with the same relay counter synchronously. The relay length T_{relay} is the time between transmissions of packets with relay counter c and $c + 1$.

Consequently, a node can infer from the relay counter how many times a received packet has been relayed. Besides the first transmission by the initiator, in Glossy nodes transmit a packet only as a consequence of a successful reception (see Figure 2.7). Therefore, nodes that receive packets with relay counter c synchronously transmit packets with the same relay counter $c + 1$.

As indicated in the lower part of Figure 2.5, we define the time between the start of a packet transmission with relay counter c and the start of the following packet transmission with relay counter $c + 1$ as the *relay length T_{relay}*. Nodes locally estimate T_{relay} using timestamps taken at the occurrence of radio interrupts. Most importantly, T_{relay} is a network-wide constant, since during a flood nodes never alter the packet length. Based on the relay counter c of a received packet and the estimate of T_{relay}, a node computes the time at which the initiator started the flood, called the *reference time t_{ref}*. In this way, all receivers synchronize relatively to the clock of the initiator. To achieve absolute time synchronization, the initiator embeds its own clock value into the flooding packet.

Current wireless systems often run two protocols in parallel: one for flooding and one for synchronization. Glossy provides both, reducing the system complexity and the risk of unintended interactions among multiple protocols.

Temporal decoupling. A time-synchronized network is useful for many purposes. Glossy benefits from it by temporally decoupling network flooding from *all* other application tasks executing on the nodes, as depicted in the upper part of Figure 2.5. In particular, nodes know the interval between two Glossy phases (e.g., by embedding the interval into packets injected by the initiator), which allows them to synchronously stall other tasks right before a flood and to resume these tasks immediately afterwards. As a result, Glossy never interferes with other activities, leading to a highly deterministic behavior during a flood. Temporal decoupling is thus another key factor to make synchronous transmissions precisely overlap.

Moreover, temporal decoupling allows the network to run other protocols or execute other tasks between two network floods. For example, an application may use a data collection protocol on top of a low-power MAC to gather sensory data, while at regular intervals Glossy takes over to disseminate commands to the nodes (e.g., to adjust the sampling rate) and to keep the nodes synchronized. Here, also the application benefits from temporal decoupling, since flooding packets never interfere with data collection packets. Such protocol interference could substantially degrade system performance, especially in terms of data yield [CKJL09].

Glossy integrates smoothly with a software system that provides primitives to decouple tasks over time, such as the slotted programming approach [FW10]. Duty-cycled networks, where all nodes wake up at the same time [BIS+08], can allocate Glossy at the beginning of the active phase, or during the sleep phase when no other communication takes place. Nevertheless, as demonstrated by our experiments in Section 2.6, Glossy needs only a few milliseconds to complete. In Chapter 3 we show that it is indeed feasible to confine multiple Glossy floods within short communication slots that are executed sequentially at all nodes.

Figure 2.6 shows a possible way of integrating Glossy with the rest of the software system on a node. An *application* that wishes to use it instructs the *scheduler* using scheduleGlossy() to run Glossy with a certain period. This period can be changed by the application at runtime (e.g., upon receiving a new interval from the initiator) by calling the same function again. Depending on the period, the scheduler starts and stops Glossy, using functions startGlossy() and stopGlossy() provided by the Glossy interface. Moreover, the scheduler notifies the application in advance via callback function glossyStarts() before it starts Glossy, giving the application the opportunity to prepare for Glossy taking over. Similarly, the scheduler notifies the application via callback function glossyFinished() after Glossy has terminated.

This *control flow* is the same for both initiator and receiver, but the *data flow* is not, as shown in the lower part of Figure 2.6. At the initiator, the

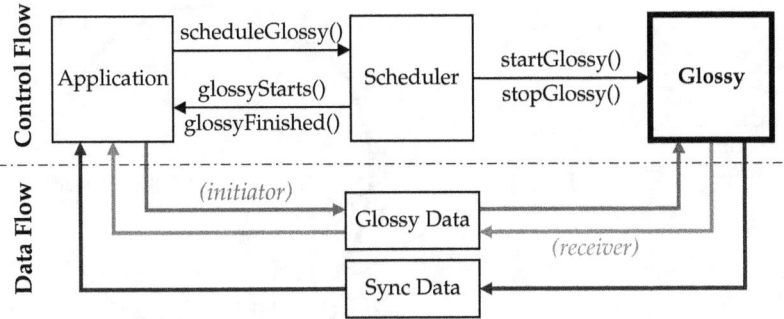

Figure 2.6: Example of Glossy as an application service. An application instructs a scheduler to run Glossy periodically. The scheduler notifies the application about an upcoming Glossy phase.

application provides Glossy with the data to be flooded. At the receiver, Glossy passes the received data to the application. Glossy provides the synchronization data (i.e., the reference time) to the applications of both, initiator and receiver.

At system startup, when a receiver is not yet synchronized, the application may instruct the scheduler to run Glossy with a shorter period in order to quickly overhear a Glossy packet from other nodes and get synchronized. By adapting the period, sophisticated mechanisms [DC08] can be implemented to achieve the desired trade-off between fast initial synchronization and energy efficiency.

Glossy manages interrupts and timers transparently. It masks software and hardware interrupts that are not essential to its functioning and disables all hardware timers. Nevertheless, Glossy records which interrupts have been active and which timers have been scheduled before its execution. Using this information, Glossy restores interrupts and timers after it terminates, allowing the application and the rest of the system to smoothly continue its execution. The application, however, is responsible for deferring actions that would be executed within the next Glossy phase.

2.3 Glossy in Detail

This section describes the Glossy architecture in detail. We first illustrate the sequence of operations executed during a flood, followed by an analysis of the timing behavior of these operations.

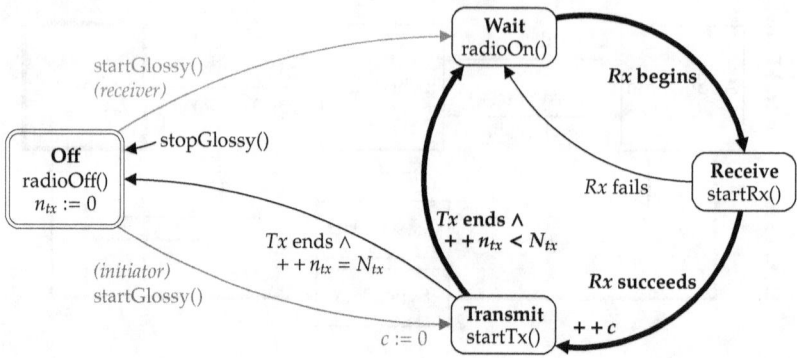

Figure 2.7: States of Glossy during execution. Transitions in the main state sequence (bold arrows) are triggered by radio events.

2.3.1   Radio-driven Execution Model

Figure 2.7 depicts the core of Glossy, represented by the repetitive sequence of states **Wait** → **Receive** → **Transmit**. The scheduler starts Glossy using startGlossy(). Afterwards, a receiver begins the execution in the **Wait** state. The initiator, instead, starts from state **Transmit**, and transmits a packet with relay counter $c = 0$. After this startup phase, the execution is the same for both initiator and receivers, as described next.

In the **Wait** state, a node has its radio turned on and waits for a packet being flooded through the network. When the radio indicates the beginning of a reception, the microcontroller unit (MCU) starts to read the incoming packet. This action corresponds to a transition to the **Receive** state. If the reception fails (e.g., due to packet corruption), the node returns to the **Wait** state and awaits a subsequent transmission from its neighbors. Otherwise, if the reception succeeds, the node makes a transition to the **Transmit** state. In this case, the MCU immediately issues a transmission request to the radio, increments the relay counter c by 1, and copies the modified packet from the receive (Rx) buffer to the transmit (Tx) buffer. To introduce only a small and predictable delay, the MCU performs this packet copying *after* issuing the transmission request, that is, while the radio switches from Rx to Tx mode. In Section 2.4, we demonstrate the feasibility of this approach on common sensor node platforms. Some recent radios feature a single packet buffer [Atm], making the packet copying step obsolete.

Nodes can transmit a packet multiple times to increase flooding reliability. We denote with N_{tx} the *maximum number of times a node transmits during a flood*. When a packet transmission ends, a *transmission counter* n_{tx} is incremented and compared to the maximum number of transmissions

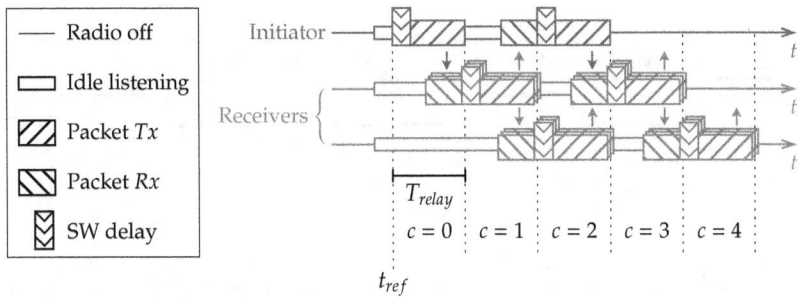

Figure 2.8: Example of a Glossy flood with $N_{tx} = 2$. Nodes always transmit packets with the same relay counter c synchronously.

N_{tx}. If a node has already transmitted N_{tx} times, it makes a transition to state **Off**, turns off the radio, and Glossy completes. Otherwise, the node returns to the **Wait** state, and the sequence starts again at the next packet reception. Ultimately, the scheduler stops Glossy by calling stopGlossy().

Figure 2.8 shows an example of a network flood with $N_{tx} = 2$. When Glossy starts, the initiator sets the relay counter c to 0 and transmits the first time to start the flooding process. Receivers within communication range of the initiator overhear the packet, set c to 1, and transmit synchronously. Their neighboring nodes, *including the initiator*, overhear this second packet, set c to 2, and again transmit synchronously. In this way, nodes always transmit packets with the same relay counter c synchronously. The process repeats until n_{tx} reaches N_{tx} at all nodes in the network. In Section 2.6, we further investigate the impact of N_{tx} on the performance of Glossy.

All transitions among states in Glossy's main loop in Figure 2.7 are triggered by radio events. On standard sensor network platforms, the MCU is typically notified of these events through interrupts. Therefore, the few software operations required by Glossy are executed within interrupt service routines (ISRs). In Section 2.4, we describe an implementation of Glossy that limits uncertainties in the execution time to inaccuracies of the underlying hardware.

An important consequence of its operation is that Glossy does not require nodes to maintain information about the network topology. For a node it is sufficient to know when a flood starts and whether it is the initiator or a receiver of such flood. In Chapter 3 we show that this property makes it possible to design efficient communication protocols that keep no topology-dependent state at the nodes, increasing their resilience to link changes due to interference, node failures, and mobility compared to prior approaches.

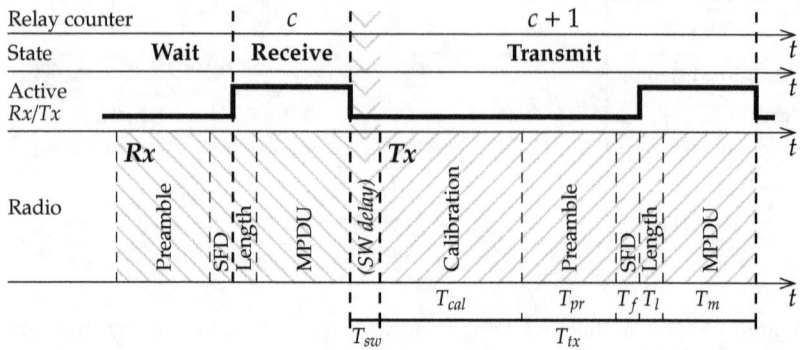

Figure 2.9: Timeline of main Glossy states. The radio determines the dwell time of each state, except for the time required to trigger a packet transmission, T_{sw}, which is determined by the MCU.

2.3.2   Execution Timing

We now show that the dwell time of the main Glossy states depends primarily on the radio hardware. The MCU influences the timing only after the completion of a packet reception, and only for the time necessary to trigger a packet transmission.

Figure 2.9 shows the timeline of the main Glossy state sequence. We see that the timing depends only on the radio, besides a short interval at the beginning of the **Transmit** state. We call this interval the *software delay* T_{sw}, required by the MCU to trigger a packet transmission. This delay depends primarily on the software routine. In addition, it is affected by the frequency of the serial peripheral interface (SPI) bus that is used on most platforms for the communication between the MCU and the radio.

In Section 2.2, we discussed that a receiver computes the synchronization reference time based on the estimate of the relay length T_{relay}, defined as the time between the start of two packet transmissions with relay counter c and $c + 1$. We now show that T_{relay} mostly depends on the radio hardware, which is an important property to achieve high synchronization accuracy. We analyze in Section 2.5 how hardware inaccuracies influence the relay length. T_{relay} accounts for the software delay T_{sw}, the time required to transmit a packet T_{tx}, and the processing delay T_d introduced by the radio at the beginning of a packet reception. T_{relay} can thus be expressed as:

$$T_{relay} = T_{sw} + T_{tx} + T_d \tag{2.1}$$

We now provide an analytical expression for the time required to transmit a packet T_{tx}. Figure 2.9 shows the operations performed by the radio during a packet transmission. Once the radio receives a transmission request, it starts to calibrate the internal voltage controlled oscillator (VCO). We denote the hardware-dependent time required for this calibration with T_{cal}. A valid IEEE 802.15.4 packet consists of the following fields: (*i*) a preamble of eight 0x0 symbols, (*ii*) the SFD corresponding to the two symbols 0x7A, (*iii*) the two-symbol frame length field that specifies the number of bytes L_m contained in the MAC protocol data unit (MPDU), and (*iv*) the MPDU itself that carries the application data. We denote with T_{pr}, T_f, T_l, and T_m the times required to transmit each of these fields. The time needed for a packet transmission is thus:

$$T_{tx} = T_{cal} + T_{pr} + T_f + T_l + T_m \tag{2.2}$$

Note that in (2.2) only T_m depends (linearly) on the packet length. The other terms are determined by the radio hardware and the standard.

2.4 Implementation

We implement Glossy using the Contiki operating system [Conb, DGV04]. We target the TelosB platform [PSC05], which features a 16-bit MSP430 MCU [Texe] and a IEEE 802.15.4-compliant CC2420 radio [Texb] supporting a transmission rate of 250 kbps. A TelosB node features 10 kB of RAM and 48 kB of program memory.

In this section we first show how our implementation leads to a highly deterministic software delay, which is a necessary condition to generate constructive interference (see Section 2.1). We then outline the approach we use to achieve high synchronization accuracy. Finally, we provide guidelines for porting Glossy to different radios.

Software instructions are executed by the MCU, whose clock is sourced by a digitally controlled oscillator (DCO). The DCO frequency varies with temperature, voltage, and from device to device. Although digital control allows to stabilize the frequency on a long-term basis (e.g., using the more stable external 32,768 Hz crystal as a reference), short-term stability is not guaranteed. The frequency of the DCO can deviate up to ±20 % from the nominal value, with temperature and voltage drifts of -0.38 %/°C and 5 %/V [Texe]. To counteract these variations, our Glossy implementation (*i*) strives to *minimize* the number of software instructions, mitigating the impact of DCO instability on the software delay, and (*ii*) ensures a *constant* number of software instructions. Moreover, the DCO runs independently of the radio clock, leading to varying delays in the transfer of digital signals between the radio circuit and the MCU. Our implementation thus (*iii*) compensates for varying offsets between the DCO and the radio clock. In the following, we describe these aspects of our implementation.

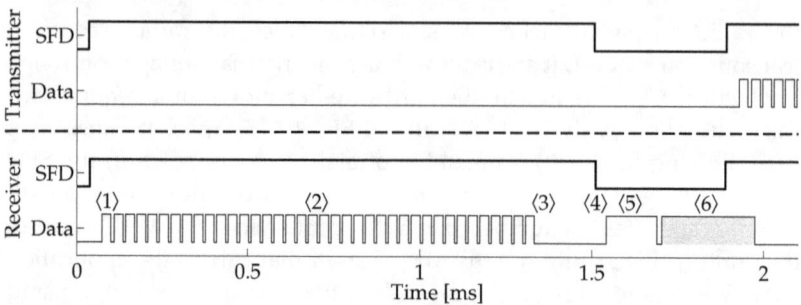

Figure 2.10: Data transfer between the radio buffers and the MCU. Snapshot taken with a logic analyzer during the transmission of a 46-byte packet. Data transfers (Data) from the radio's Rx buffer are in white, to the Tx buffer in gray.

2.4.1 Minimizing Software Delay

We exploit buffered packet receptions and transmissions to minimize the number of software instructions. The CC2420 radio provides two buffers for receiving (Rx) and transmitting (Tx), implemented as first-in-first-out (FIFO) queues. During a packet reception, the radio stores incoming data into the Rx buffer. The FIFO pin signals the MCU when the Rx buffer contains at least 1 byte. The MCU never enters a low-power mode while Glossy is running, so that all of its components are always enabled.

Glossy can flood packets of any IEEE 802.15.4-compliant length. During a reception, a receiver checks the length of the packet being flooded by reading the first byte of the packet, corresponding to the packet length field. This action corresponds to step ⟨1⟩ in Figure 2.10, which shows the sequence of data transfers between the radio buffers and the MCU measured with a logic analyzer. If the packet length is greater than 8 bytes, Glossy lets the MCU poll the FIFO pin: the content of the packet is read byte after byte over the SPI bus and stored into a temporary buffer while the packet is still being received by the radio ⟨2⟩. When only 8 bytes are left to receive, Glossy stops polling the FIFO pin ⟨3⟩, and waits until the SFD pin makes a transition to 0 and an interrupt occurs. At this point, Glossy executes a minimum, constant number of software instructions just to serve the interrupt and to issue a transmission request to the radio ⟨4⟩. Then, while the radio calibrates the VCO, Glossy reads the last 8 bytes ⟨5⟩ from the Rx buffer, and checks whether the packet has been received successfully. If so, Glossy increments the relay counter c and copies the data from the temporary buffer into the Tx buffer of the radio ⟨6⟩, which then transmits the packet; otherwise, Glossy aborts the transmission process before the radio actually starts sending the preamble. When the packet length is at most 8 bytes, Glossy skips polling the FIFO pin (i.e., steps $\langle 2 \rangle$ and $\langle 3 \rangle$), and reads the remaining content of the Rx buffer after the packet reception completes.

This approach is feasible as it takes only a few software instructions to check whether a packet has been successfully received; the MCU executes these instructions and starts writing to the Tx buffer before the radio completes the VCO calibration. For long packets, the Tx buffer might not be completely filled before the radio starts transmitting the first bytes from the Tx buffer, as is the case in Figure 2.10. This does not cause a buffer underflow, since the Tx buffer is a FIFO queue and copying data over the SPI bus is faster than transmitting it over the wireless medium. In our implementation, we measure that the latency required for writing to the Tx buffer is 5.75 μs per byte, plus an initial latency of 18 μs; the time required to transmit a byte with an IEEE 802.15.4-compliant radio is 32 μs.

2.4.2   Approaching a Deterministic Software Delay

A small and constant number of software instructions does not guarantee that the MCU issues a transmission request to the radio a constant number of cycles after being notified of the end of a packet reception. When the MCU receives an interrupt request, it first completes the execution of the current instruction before starting to serve the interrupt. Instructions on the MSP430 require between 1 and 6 clock cycles to complete [Texe]. Interrupts are thus served with a variable delay, depending on which instruction is being executed when the MCU receives the interrupt.

Glossy compensates for this variable delay by measuring the number of clock ticks elapsed between the instant at which the interrupt is received, recorded using the capture functionality of the MCU at the falling edge of the SFD pin, and the instant at which it is served. Depending on the measured delay, Glossy inserts a certain number of *no operations (NOPs)* at the beginning of the interrupt handler. Glossy thus ensures that a transmission request to the radio is issued a constant number of clock cycles after the interrupt reception, which makes the software delay highly deterministic.

2.4.3   Compensating for Hardware Variations

The software delay T_{sw} is the sum of (*i*) the time required by the MCU to sample the transition of the radio's SFD pin at the end of a packet reception, (*ii*) the number of MCU clock ticks I needed to issue a transmission request to the radio, and (*iii*) the time required by the radio to sample the transmission request coming from the MCU. Using the mechanisms described above, we know that I is constant in our implementation. However, due to the asynchronous clocks of radio and

MCU, the software delay T_{sw} is still not constant: there is a variable delay in the transfer of digital signals between these two components.

The radio updates its digital output signals with frequency f_r, determined by its crystal oscillator. The internal DCO determines the frequency f_p of the MCU. Neglecting clock drifts, we can write:

$$T_{sw} = \frac{1}{f_r} \times \left\lceil (I + k_p) \times \frac{f_r}{f_p} \right\rceil \qquad (2.3)$$

where k_p ($0 < k_p \le 1$) is the fraction of the DCO period $1/f_p$ required at the MCU to sample the SFD transition at the end of the packet reception. Given that the radio clock and the DCO run completely unsynchronized, the initial offset k_p is a continuous random variable uniformly distributed in the interval $0 < k_p \le 1$. This implies that T_{sw} is a discrete random variable with granularity $1/f_r$. The number of possible discrete values for T_{sw} and their distribution depend on the number of DCO ticks I.

By inserting a constant number of NOPs, we choose I in our implementation so that we obtain from (2.3) a distribution that achieves the theoretical lower bound of only two possible values for T_{sw}. These two values are $1/f_r$ apart—as long as radio and MCU use two independent clocks, any implementation of Glossy exhibits a minimum jitter of $1/f_r$ in the software delay. For example, on the TelosB, the CC2420 radio updates its digital output signals with frequency $f_r = 8$ MHz, and the DCO of the MSP430 runs in our implementation at a frequency $f_p = 4,194,304$ Hz. The resulting difference between the two possible values of T_{sw} is 125 ns.

We measured the software delay of our implementation by connecting the radio SFD pin of four TelosB nodes to a logic analyzer. These four nodes act as receivers, while an additional node periodically initiates a flood. Upon overhearing a packet from the initiator, the receivers trigger a transmission, and we measure the distance between the end of a reception and the beginning of a transmission. This corresponds to $T_{sw}+T_{cal}+T_{pr}+T_f$.

Figure 2.11 shows the distribution of the software delay, computed by removing the constant $T_{cal} + T_{pr} + T_f = 352$ µs. We see that in 91 % of the cases the delay matches the theoretical binary distribution according to (2.3). Moreover, the values are equally distributed between the two possible values, 23.25 µs and 23.375 µs, for $I = 97$ in our implementation. In the remaining 9 % of the cases, an additional uncertainty of one radio clock tick affects the delay. However, this is mainly due to the drift of the DCO that may generate a frequency different from the nominal value f_p. In fact, it appears very difficult to avoid such negative drift effects, given that our implementation synchronizes the DCO with respect to the stable 32,768 Hz crystal every time Glossy starts, and the number of instructions is (at least very close to) the minimum.

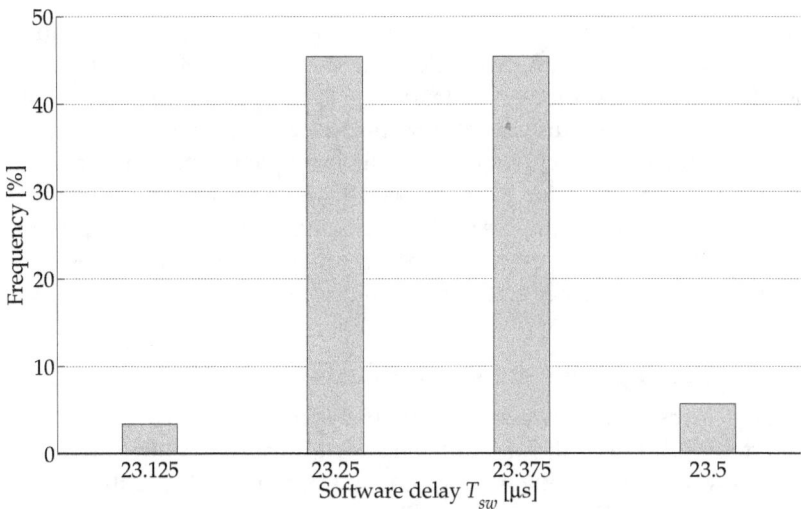

Figure 2.11: Distribution of the software delay T_{sw}. Results from a logic analyzer show that in 91 % of the cases T_{sw} matches the theoretical binary distribution of two values with a distance of one radio clock tick, corresponding to 125 ns.

2.4.4   Time Synchronization

Glossy provides implicit time synchronization. During a flood, receivers compute a common reference time t_{ref} based on the relay counter c received during the flood and the estimated relay length T_{relay}. On a TelosB sensor node, the MSP430 can use two separate time sources: the internal high-frequency DCO and an external low-frequency crystal. This 32 kHz external crystal is significantly more stable than the internal DCO but provides time with low resolution. By contrast, the DCO provides sub-microsecond resolution, but does not guarantee short-term stability, and is usually disabled when the MCU enters low-power execution modes.

We achieve high-resolution and low-power time synchronization by employing *Virtual High-resolution Time* (VHT) by Schmid et al. [SDS10]. The DCO is enabled at the beginning of a Glossy phase, and the MCU does not enter a low-power mode until a Glossy phase terminates. The MCU timestamps with the high-frequency clock all the interrupts generated by transitions of the SFD pin. As a result, receivers compute high-resolution estimates of T_{relay}. The timer capture functionality of the MSP430 is then exploited to translate the high-resolution estimate of the reference time t_{ref} to a low-resolution value and a relative high-resolution offset. At the end of a Glossy phase, these two time values are provided to the application.

When the application schedules synchronized actions, it only needs to turn on the internal DCO and do a reverse translation to a high-resolution time value. As a result, events can be scheduled with high resolution and with an energy cost proportional to the number of timer accesses. Receivers can also exploit estimates of T_{relay} to compute the drift of the low-frequency clock. As discussed in Section 2.3.2, T_{relay} depends to a great extent on the radio clock, sourced by a high-frequency and stable crystal. Accurate synchronization between two Glossy phases can be maintained by compensating for the measured drift. We show in Section 2.6 that our synchronization implementation accurately estimates the reference time t_{ref} with an average error smaller than one microsecond.

2.4.5   Porting Glossy to Other Radios

Glossy can be ported to other IEEE 802.15.4-compliant radios. Destructive interference due to path delay differences is not a major problem in current low-power wireless networks, where links are rarely longer than a few tens of meters [DDHC+10]. With long-range radios, the requirement for constructive interference $\Delta_{max} = 0.5\,\mu s$ corresponds to a maximum *difference* in path delay of 150 meters. However, if transmission power control is not used, such big differences in path delay would also result in big differences in received signal strength, making a correct reception of the first (stronger) packet very likely due to capture effects.

Radios like the CC2520 [Texc] feature a RAM for packet buffering. It is thus possible to change the value of certain bytes (e.g., the relay counter) in-situ, that is, without the need to transfer the entire packet twice over the SPI bus between a reception and a transmission. Compared to our TelosB implementation, this feature eases the effort for minimizing the software delay. Moreover, such delay is completely eliminated in radios providing automatic switch to transmission mode at the end of a reception [Texa].

Kuo et al. have recently released μSDR, a low-cost, low-power, portable software-defined radio (SDR) platform that allows to develop timing-critical communication protocols for low-power wireless networks [KPSD12]. μSDR natively supports Glossy in hardware, sparing the need to disable unnecessary interrupts during a flood and to execute software routines at each packet relay. Moreover, by using μSDR it is possible to tune some low-level radio parameters (e.g., automatic gain control and carrier frequency) and further improve the performance of Glossy compared to commercial radios [KPSD12].

More recently, Carlson et al. [CCT+13] have successfully ported Glossy to the CC430 platform [Texd], which combines on a single chip an MSP430 MCU with a radio operating on bands below 1 GHz. It is important to notice that the latter is not compliant to the IEEE 802.15.4 standard, which demonstrates that the Glossy architecture can support diverse wireless bands, modulation schemes, and data rates. In their implementation, for

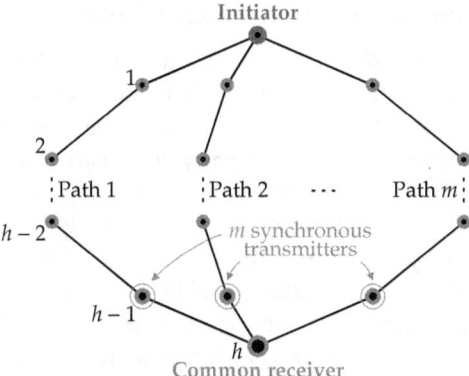

Figure 2.12: Scenario for the theoretical analysis: a receiver is at the end of m independent paths of length h.

example, the CC430 operates on the 900 MHz band, using a 2-Frequency-Shift Keying (2-FSK) modulation scheme and a 125 kbps data rate.

2.5   Theoretical Analysis

This section studies Glossy analytically. In particular, we look at the probability that Glossy makes synchronous transmissions of the same packet interfere constructively, which depends on the temporal displacement Δ among the transmissions (see Section 2.1). We analyze how Δ is affected by the number of synchronous transmitters (i.e., node density) and the maximum hop distance of a receiver from the initiator (i.e., network diameter). In addition, we study the limits of Glossy in worst-case settings that are difficult to reproduce in real networks.

To this end, we consider the network structure in Figure 2.12. There are $m \geq 2$ *independent* flooding paths originating at the initiator. These paths traverse $h \geq 2$ hops each, and join again at a common receiver. In this way, we construct a worst-case scenario in the sense that the initiator provides the only common synchronization point to the paths: nodes on one path relay the flooding packet independently of nodes on the other paths, which challenges Glossy in making the final m synchronous transmissions interfere constructively at the common receiver.

We first present a theoretical model of the temporal displacement Δ experienced by the common receiver, independent of a specific implementation. Then, we apply this model to our implementation on TelosB devices. Results show that constructive interference occurs with a probability above 99.9 % even when 30 nodes transmit synchronously.

2.5.1   Implementation-Independent Analysis

We first analyze the sources of temporal uncertainty affecting the relay length T_{relay} given by (2.1). Our analysis makes use of a mixture of statistical and worst-case assumptions. We consider statistical distributions for processes that are clearly stochastic in nature, such as the offset between two unsynchronized clocks. We instead consider worst-case scenarios for more deterministic variables, including clock drift, network topology, and the maximum temporal displacement among transmitters. Hence, our model provides the statistical worst-case displacement Δ experienced by the common receiver as a function of m and h. This model can then be used to compute the theoretical reliability and synchronization accuracy for a specific network scenario.

2.5.1.1   Statistical Uncertainty on Relay Length

We discussed in Section 2.4 that the delay T_{sw} introduced by the software routine to trigger a transmission is in general not constant, even if the number of instructions I executed by the MCU is fixed. The software delay T_{sw} is a multiple of the period $1/f_r$ of the crystal sourcing the radio clock. We can thus express the software delay as $T_{sw} = \tilde{T}_{sw} + \tau_{sw}$, where $\tilde{T}_{sw}$ is a constant value corresponding to the minimum possible delay, and τ_{sw} is a discrete random variable with granularity $1/f_r$ representing the additional variation due to the unsynchronized clocks of the MCU and the radio. We denote with p_{sw} the probability mass function (pmf) of τ_{sw}.

The processing delay of the radio, T_d, is also not constant. The digital circuits of the radio are sourced by a crystal oscillator that has frequency f_r. The radio starts to process an incoming packet when the digital circuits sampled the beginning of a reception, at most after $1/f_r$. We express the processing delay of the radio as $T_d = \tilde{T}_d + \tau_d$. The time needed to process an incoming packet, $\tilde{T}_d$, is a constant usually in the order of a few microseconds and determined by the radio. The time required to sample a reception, τ_d, depends on the offset between the radio clocks of the transmitter and the receiver. Since these clocks run unsynchronized, τ_d is a random variable with uniform distribution in the interval $[0, 1/f_r]$. For simplicity, we discretize the set of values of τ_d by introducing a time granularity δ such that $\delta = 1/(k \times f_r)$, with $k \gg 1$. Consequently, τ_d has uniform discrete distribution $\tau_d = \{0, \delta, 2 \times \delta, \ldots, 1/f_r\}$ and pmf p_d with constant values $1/(k + 1)$.

The statistical uncertainty on the length of a relay is the sum of the uncertainties on the software delay and the radio processing:

$$\tau_{relay} = \tau_{sw} + \tau_d \tag{2.4}$$

Since $\delta \ll 1/f_r$, τ_{relay} has granularity δ. The two uncertainties, τ_{sw} and τ_d, are independent, because they are due to independent effects. Recalling that the distribution of the sum of two independent random variables can be expressed by their convolution, the uncertainty on the length of a relay has pmf $p_{relay} = p_{sw} * p_d$.

2.5.1.2   Worst-Case Drift of Radio Clock

The time required for a packet transmission, T_{tx}, is given by (2.2) and depends on the frequency of the radio clock f_r. In general, this frequency deviates from the nominal value $\tilde{f}_r$ due to temperature and aging effects. Crystals used to source the clocks of sensor network radios typically have a frequency drift ρ that depends on the temperature t according to a third-order polynomial [Sch09]:

$$\rho = (f_r - \tilde{f}_r)/\tilde{f}_r = A(t - t_0)^3 + B(t - t_0) + C \tag{2.5}$$

Here, A, B, C, and t_0 are constants that depend on the specific crystal device. Using (2.5), it is possible to determine bounds on the frequency drift for a given temperature range.

In the following, we assume $-\overline{\rho} \le \rho \le \overline{\rho}$. In the worst case, among the m independent paths in Figure 2.12, there is at least one path where *all* radio clocks run at the highest frequency $\tilde{f}_r \times (1 + \overline{\rho})$, and at least one other path where *all* radio clocks run at the lowest frequency $\tilde{f}_r \times (1 - \overline{\rho})$. We denote with $\tilde{T}_{tx}$ the nominal transmission time in the absence of radio clock drift. With this, we can express the worst-case variation on the transmission time that accumulates after h hops at the end of these two independent paths as follows:

$$\tau_{tx} = (h - 1) \times \frac{\tilde{T}_{tx}}{1 + \overline{\rho}} + (h - 1) \times \frac{\tilde{T}_{tx}}{1 - \overline{\rho}}$$
$$= (h - 1) \times \tilde{T}_{tx} \times \frac{2 \times \overline{\rho}}{1 - \overline{\rho}^2} \tag{2.6}$$

2.5.1.3   Statistical Worst-Case Temporal Displacement

Each node introduces a statistical uncertainty τ_{relay} on the length of a relay. This uncertainty is independent of other nodes, since the pair of radio and MCU clocks on one node runs independently from the pair of clocks on other nodes. Therefore, the temporal uncertainty τ associated with a path that consists of h hops is the sum of $h - 1$ independent random variables:

$$\tau = (h - 1) \times \tau_{relay}$$
$$= (h - 1) \times (\tau_{sw} + \tau_d) \tag{2.7}$$

The pmf of τ is given by the convolution of $h - 1$ instances of p_{relay}.

We now extend the problem to m independent paths, each consisting of h hops and originating at the initiator. We are interested in the statistical worst-case temporal displacement Δ. This displacement corresponds to the difference between the maximum and the minimum timing variation associated with each path.

We consider a worst-case scenario where the path with the minimum time variation has clocks running at the highest frequency, and the path with the maximum variation has clocks running at the lowest frequency:

$$\Delta = \max_m[\tau] - \min_m[\tau] + \tau_{tx} \qquad (2.8)$$

Based on (2.8), we want to determine the cumulative distribution function (CDF) of Δ. This corresponds to the problem of finding the CDF of the sample range of m independent identically distributed (i.i.d.) experiments. This is a well-known order statistics problem, and analytical solutions exist in the literature [ABN08].

2.5.2   Implementation-Dependent Analysis

We now apply the above model to our Glossy implementation on TelosB devices. We analyze the statistical worst-case temporal displacement Δ for different node densities and network diameters. We use the measurements shown in Figure 2.11 to obtain the pmf of the software delay T_{sw}. In addition, we consider $\bar{\rho} = 20$ parts per million (ppm) as the maximum drift affecting the radio crystal, which corresponds to a temperature range between -30°C and 50°C [Sch09].

To analyze the dependence on network diameter, we fix the number of paths at $m = 2$ and vary the path length h between 2 and 8 hops. The maximum number of transmissions is set to $N_{tx} = 3$. Figure 2.13(a) plots the CDF of Δ for four different settings. We see that Δ is smaller than the requirement of 0.5 μs with very high probability. This shows that when a flooding packet is relayed along two independent paths over 8 hops, the two final synchronous transmissions generate constructive interference in more than 96 % of the cases. Recall that this assumes that the time variation due to radio clock drift is maximum between the two paths. In most cases, however, the drift is much smaller than its bounds, which increases the probability of constructive interference significantly.

To analyze the dependence on node density, we set the path length to $h = 2$ and vary the number of paths m between 2 and 30. Figure 2.13(b) shows that our Glossy implementation is robust also in dense networks. In fact, when 30 nodes transmit synchronously to a common receiver, Δ is below 0.5 μs with a probability above 99.9 %. We show in Section 2.6 that experiments on high-density networks confirm these analytical results.

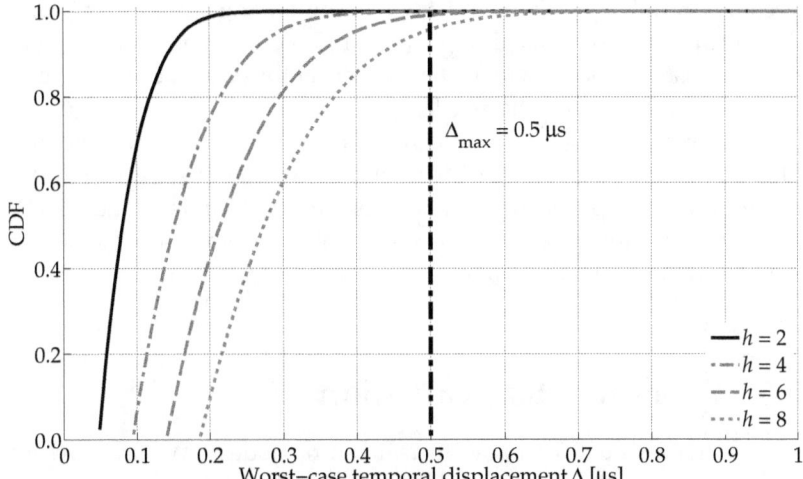

(a) Distribution of Δ as a function of network diameter h, with $m = 2$.

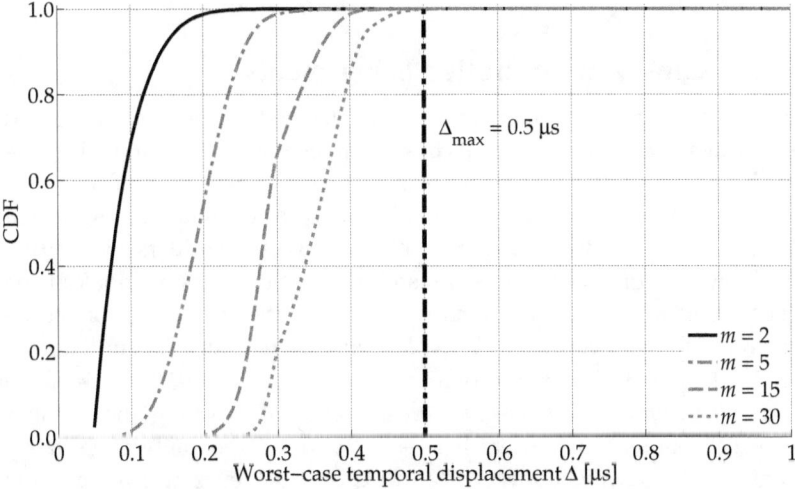

(b) Distribution of Δ as a function of node density m, with $h = 2$.

Figure 2.13: Results of the theoretical analysis. In our Glossy implementation, the statistical worst-case temporal displacement Δ among all transmitters is smaller than $0.5\ \mu s$ with different network settings. Results are for $N_{tx} = 3$.

2.5.3   Theoretical Lower Latency Bound

We provide an expression for the theoretical lower bound on flooding latency. Intuitively, this bound is given by adding up the hardware-dependent times for transmitting T_{tx} and radio processing T_d. Given that nodes transmit synchronously, the theoretical lower latency bound in a network with diameter h hops is $h \times (T_{tx} + T_d)$.

At each hop, Glossy adds a delay T_{sw} to the theoretical lower bound. This delay is introduced by the MCU when issuing a transmission request to the radio. In our implementation, T_{sw} is at most 23.5 μs (see Figure 2.11). We are not aware of any flooding protocol that comes so close to the theoretical lower latency bound.

2.6   Experimental Evaluation

This section evaluates Glossy on real sensor nodes. We first present results from experiments with a few nodes in several controlled settings. Afterwards, we report on the performance of Glossy during extensive experiments on three wireless sensor testbeds.

2.6.1   Glossy in Controlled Experiments

Before evaluating Glossy on several testbeds, we use controlled experiments to study some of its basic characteristics. We start by looking at the *reliability of synchronous transmissions*, defined as the fraction of packets correctly received by a node. We analyze how reliability is affected by the temporal displacement Δ between two transmitters and by the total number of synchronous transmitters. Afterwards, we look at the *time synchronization error*, which we define as the absolute error on the reference time t_{ref} computed by a receiver with respect to the initiator.

We find that (*i*) Glossy provides a reliability above 95 % *in a scenario where the capture effect does not occur*; (*ii*) while varying the number of synchronous transmitters between 2 and 10, reliability stays fairly constant and always above 98 %; (*iii*) Glossy achieves an average time synchronization error of less than 0.4 μs even at receivers that are eight hops away from the initiator.

2.6.1.1   Impact of Temporal Displacement

The first experiment evaluates the reliability of Glossy in a scenario where the capture effect does not occur. In this case, a successful reception is only possible if synchronous transmissions interfere constructively. While this is clearly a worst-case scenario that is difficult to reproduce even under controlled settings, it provides an indication of the robustness of Glossy.

Setup. We use three nodes, one initiator and two receivers, and set $N_{tx} = 1$. Upon receiving a packet from the initiator, the two receivers transmit synchronously. The initiator overhears these transmissions and records whether it can successfully decode the packet. Based on sequence numbers embedded in the packets, we measure the reliability experienced by the initiator. Moreover, we delay the transmission of one receiver by a variable amount of time in the interval $[0, 8]$ μs by letting the receiver execute a certain number of NOPs before issuing a transmission request to the radio. We set the clock frequency of the MCU to 4 MHz to obtain a temporal displacement Δ with 250 ns granularity between the two receivers, corresponding to half-chip period $T_c/2$ of the modulated signal.

To ensure that the capture effect does not help the initiator in receiving the packet, we adjust the transmit powers of the receivers. In particular, we let the non-delayed receiver transmit at -20 dBm, and the delayed receiver at -13 dBm. In this way, since both signals are weak and the second one is stronger than the first one, we prevent the initiator from capturing the first of the two signals.

Results. Figure 2.14(a) shows reliability for different temporal displacements Δ and packet lengths, averaged over 2,000 packets for each setting. We see that the leftmost bar, corresponding to Glossy without artificial delay, indicates a reliability above 95 % for short packets. This demonstrates that Glossy makes synchronous transmissions interfere constructively, allowing a receiver to decode a packet with very high probability even in the absence of the capture effect. For increasing Δ we see a pattern similar to the one in Figure 2.3 obtained through simulation. In particular, reliability starts to drop significantly at $\Delta_{max} = 0.5$ μs, showing local minima when different chips perfectly overlap. Finally, similar to non-synchronous transmissions [SAM03], reliability decreases as packets become longer.

Figure 2.14(b) shows average correlation of the first 8 symbols received after the SFD. This 7-bit value represents a measurement of the chip error rate, and is automatically appended by the radio to each successfully received packet. A value close to 110 indicates that a packet was received with maximum quality, while a value of 50 is typically the minimum required to successfully receive a packet [Texb]. We see that average correlation is maximum when no artificial delay is added to one of the two synchronous transmitters, while it drops significantly when the temporal displacement Δ equals or exceeds $\Delta_{max} = 0.5$ μs.

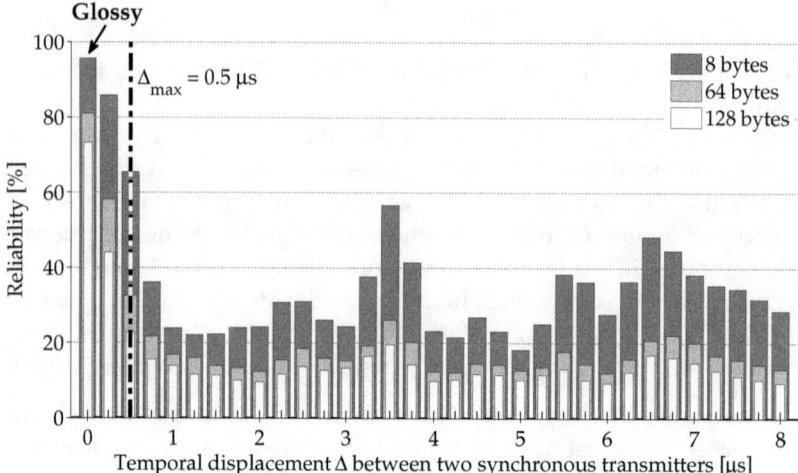

(a) Reliability is maximum when $\Delta = 0$, but drops significantly when Δ equals or exceeds $\Delta_{max} = 0.5\,\mu s$, following a pattern similar to that in Figure 2.3.

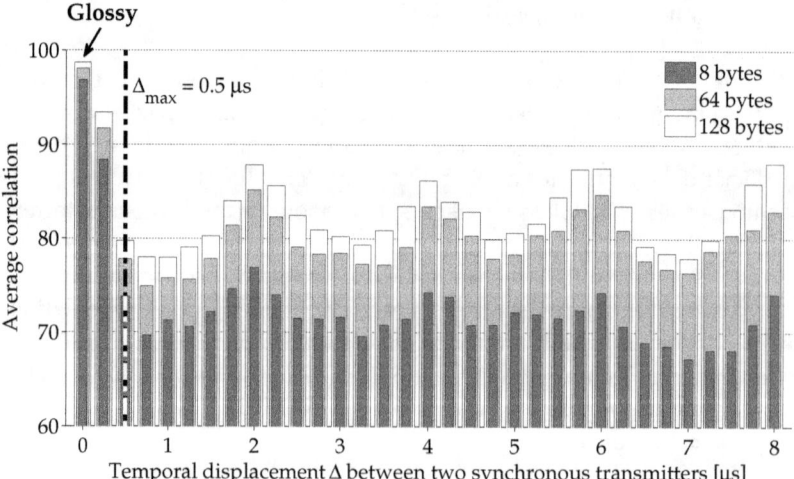

(b) Average correlation is maximum when $\Delta = 0$, but drops significantly when Δ equals or exceeds $\Delta_{max} = 0.5\,\mu s$.

Figure 2.14: Glossy in a scenario without capture effects, for $N_{tx} = 1$. Due to constructive interference, reliability and average correlation are maximum when the temporal displacement Δ is zero.

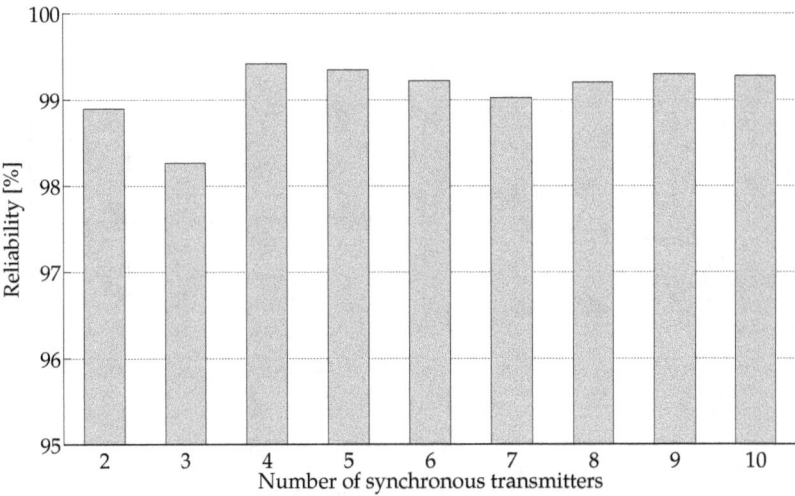

Figure 2.15: Reliability depending on number of synchronous transmitters, including capture effects, for $N_{tx} = 1$. Reliability is always above 98 % and shows no significant dependence on the number of synchronous transmitters.

2.6.1.2   Impact of Number of synchronous Transmitters

In a typical deployment, nodes are not evenly distributed and experience different channel characteristics. As a result, some nodes have more neighbors than others. We therefore study in this experiment the impact of the number of synchronous transmitters on reliability.

Setup. We use a setup similar to the one above. However, we vary the number of receivers between 2 and 10, and do not delay their transmissions artificially; all packets are 8 bytes long. In this way, we measure the reliability experienced by the initiator for different numbers of synchronous transmitters, including capture effects.

Results. Figure 2.15 shows reliability, averaged over 10,000 packets for each setting. We see that reliability stays fairly constant and always above 98 % as the number of transmitters increases, thus showing no significant dependence between the two. Interestingly, reliability is slightly lower when only two or three nodes transmit synchronously. In these settings it is more likely that the initiator receives a weak signal, due to the generation of carriers with slightly different frequencies or phases. Our results resemble those in [DMEST08], where nodes transmit fixed-length acknowledgment packets automatically generated by the radio hardware. By contrast, Glossy transmits variable-length packets generated in software, achieving even higher reliability in some cases.

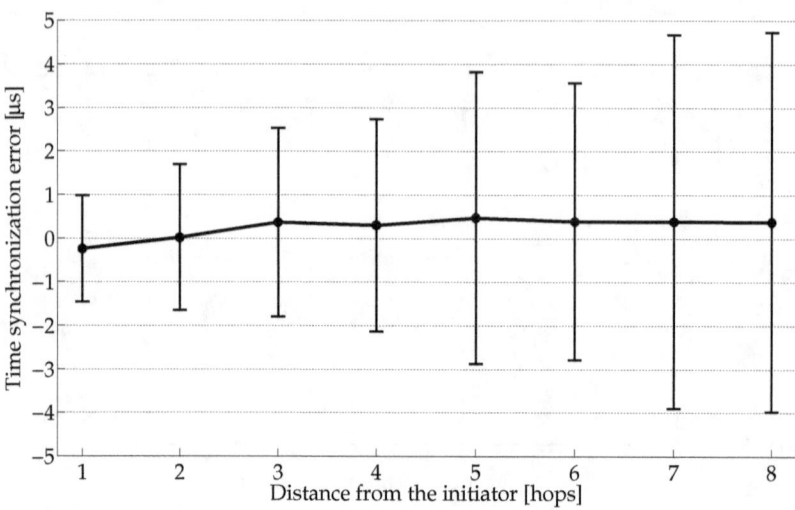

Figure 2.16: Accuracy of time synchronization in Glossy. The absolute error on the reference time t_{ref} computed by a receiver is below 0.4 µs, even at receivers that are 8 hops apart from the initiator.

2.6.1.3   Accuracy of Time Synchronization

Glossy provides network-wide time synchronization at no additional cost. To this end, a receiver estimates the reference time t_{ref} when receiving the first flooding packet, as detailed in Section 2.4.4. We assess the accuracy of this computation by looking at the absolute error with respect to the reference time t_{ref} computed by the initiator.

Setup. We use five nodes, one initiator and four receivers. At the beginning of a Glossy phase, the initiator sends a packet, computes the reference time t_{ref}, and schedules the next phase based on this reference time. The receivers do exactly the same. After receiving the packet from the initiator, they compute reference times t_{ref}, and use these to schedule the beginning of the next phase.

All nodes activate an external GPIO pin when a phase starts. We connect the respective GPIO pin of the five nodes to an oscilloscope, monitoring the start of a phase at a granularity of 2 ns. Then we measure the time difference between pin activation at the initiator and pin activation at the receivers. In this way, we obtain four estimates of the time synchronization error. To analyze this error for receivers that are more than one hop away from the initiator, we choose a sufficiently large N_{tx} and let the receivers compute the reference time based on packets received after been relayed multiple times.

Results. Figure 2.16 shows average and standard deviation of the time synchronization error depending on hop distance from the initiator, averaged over 4,000 measurements for each setting. We see that the average error remains as low as 0.4 μs up to hop 8. The standard deviation increases almost linearly with hop distance, reaching 4.8 μs at hop 8.

These results show that Glossy achieves accurate time synchronization also at receivers that are several hops away from the initiator. Moreover, they confirm a major result of our theoretical analysis in Section 2.5, namely that Glossy accumulates only a very small timing error at each relay. Most of the error is indeed of a stochastic nature, independent across nodes and different floods.

2.6.2   Glossy in Testbed Experiments

Using experiments on three wireless sensor testbeds, we evaluate Glossy's performance across several node densities, network diameters, packet lengths, and transmit powers. Results demonstrate that Glossy provides robust and efficient network flooding under diverse conditions. We first describe the testbeds and metrics we use. Then we summarize our key findings, followed by a detailed discussion of the experimental results.

2.6.2.1   Scenario and Metrics

We use three different testbeds to evaluate Glossy: MoteLab [WASW05], Twist [HKWW06], and DSN [DBK⁺07]. These differ along several aspects, including number of nodes, node density, and network diameter.

- On MoteLab, we collect data from 94 nodes unevenly spread over three floors. A node at the corner of the second floor acts as initiator. It reaches all other nodes within at most 5 hops when nodes transmit at the highest power setting of 0 dBm. When transmitting at -7 dBm, the lowest power that keeps the network fully connected, the farthest nodes are 8 hops away from the initiator.

- On Twist, we use 92 nodes and randomly choose one of them as initiator. Due to its high node density, the network stays connected even at a transmit power of -25 dBm, yielding a maximum hop distance of 5 from the initiator.

- The DSN testbed consists of 39 nodes distributed in several offices, passages, and storerooms; two nodes are located outside on the rooftop. The initiator reaches all nodes within 7 hops at the lowest possible transmit power of -15 dBm.

Testbed		MoteLab (94 nodes)		Twist (92 nodes)			DSN (39 nodes)	
Tx power [dBm]		0	-7	0	-15	-25	0	-15
Diameter [hops]		5	8	3	3	5	3	7
$N_{tx} = 1$	R [%]	99.37	94.80	99.90	99.83	99.64	99.71	98.25
	L [ms]	1.77	2.28	0.81	1.18	1.74	1.06	1.81
	T_{on} [ms]	3.13	4.85	1.99	2.39	3.04	2.31	3.45
$N_{tx} = 2$	R [%]	99.88	99.09	>99.99	99.99	99.97	99.97	99.91
	L [ms]	1.79	2.35	0.81	1.18	1.75	1.07	1.83
	T_{on} [ms]	4.75	5.14	3.37	3.81	4.56	3.76	4.75
$N_{tx} = 3$	R [%]	99.96	99.78	>99.99	>99.99	>99.99	>99.99	99.99
	L [ms]	1.79	2.37	0.81	1.18	1.75	1.07	1.83
	T_{on} [ms]	6.30	6.31	4.76	5.25	6.14	5.20	6.26
$N_{tx} = 6$	R [%]	>99.99	99.98	>99.99	>99.99	>99.99	>99.99	>99.99
	L [ms]	1.79	2.39	0.81	1.18	1.75	1.07	1.83
	T_{on} [ms]	10.87	10.18	9.07	9.60	10.84	9.52	10.79

Table 2.1: Testbed configurations and results when Glossy floods 8-byte packets. *Tx* power is the transmit power of all nodes in a testbed, and diameter is the corresponding maximum hop distance between initiator and receivers. The table shows network-wide averages of flooding latency *L*, flooding reliability *R*, and radio on-time T_{on} across four different choices of N_{tx}, which is the maximum number of transmissions per node during a network flood.

On all three testbeds, we use channel 26 to limit the interference with co-located Wi-Fi networks. The upper part of Table 2.1 lists number of nodes, transmit powers, and network diameters of each testbed we use.

   Our evaluation is based on the following three metrics:

- *Flooding latency L* of a receiver is the time between the first transmission at the initiator and the first successful reception of the flooding packet at that receiver.

- *Flooding reliability R* is the fraction of network floods in which a receiver successfully receives the flooding packet.

- *Radio on-time T_{on}* is the time a receiver has its radio turned on during a network flood.

We compute these metrics for each particular setting based on 50,000 network floods. We report *L*, *R*, and T_{on} for each receiver individually as well as averaged over all receivers in a testbed.

2.6.2.2   Summary of Testbed Results

Table 2.1 summarizes the results collected on the three testbeds when Glossy floods 8-byte packets. It lists network-wide averages of L, R, and T_{on} across four choices of N_{tx}. Based on our experiments, we find that:

- The empirical performance of Glossy exhibits no noticeable dependence on node density. Theoretically, performance and node density are not independent. The analysis in Section 2.5 shows that the probability of constructive interference is 99.9 % when adding 30 synchronous transmitters. However, we do not discern this marginal difference in our results from controlled experiments with up to 10 transmitters (see Section 2.6.1.2) and experiments on three testbeds, including TWIST where nodes are densely deployed.

- The performance of Glossy depends on network diameter, that is, on the maximum hop distance between initiator and receivers. As the network diameter increases, flooding latency L and radio on-time T_{on} increase linearly, while flooding reliability R decreases. Nevertheless, for MoteLab's largest diameter of 8 hops and $N_{tx} = 6$, L averages 2.4 ms, R is above 99.9 %, and T_{on} is as low as 10.2 ms.

- Increasing the maximum number of transmissions N_{tx} significantly enhances flooding reliability. For $N_{tx} = 3$, we observe that R is at least 99.9 % across all testbeds and transmit powers except one. Since R is already very high for $N_{tx} = 1$, further increasing N_{tx} has no noticeable effect on the average flooding latency. Radio on-time increases linearly with N_{tx}, averaging about 16 ms for the highest value of $N_{tx} = 10$ in our experiments.

2.6.2.3   Impact of Packet Length

We start by evaluating how the packet length affects the performance of Glossy. To this end, we run experiments on MoteLab with three different packet lengths: 8, 64, and 128 bytes. The latter setting corresponds to the maximum packet length supported by IEEE 802.15.4-compliant radios. All nodes transmit at the maximum power of 0 dBm.

Figure 2.17 shows the resulting average Glossy performance. We see in Figure 2.17(a) that the average flooding reliability decreases as the packet length increases. For $N_{tx} = 4$, for example, R is 99.98 % when Glossy floods 8-byte packets, but it decreases to 99.90 % and 99.47 % when it floods packets with a length of 64 and 128 bytes, respectively. This is expected as current radios make decisions about the correctness of each individual symbol—a packet is accepted only if *all* symbols are correctly received. Figure 2.17(b) and Figure 2.17(c) confirm that flooding latency and radio

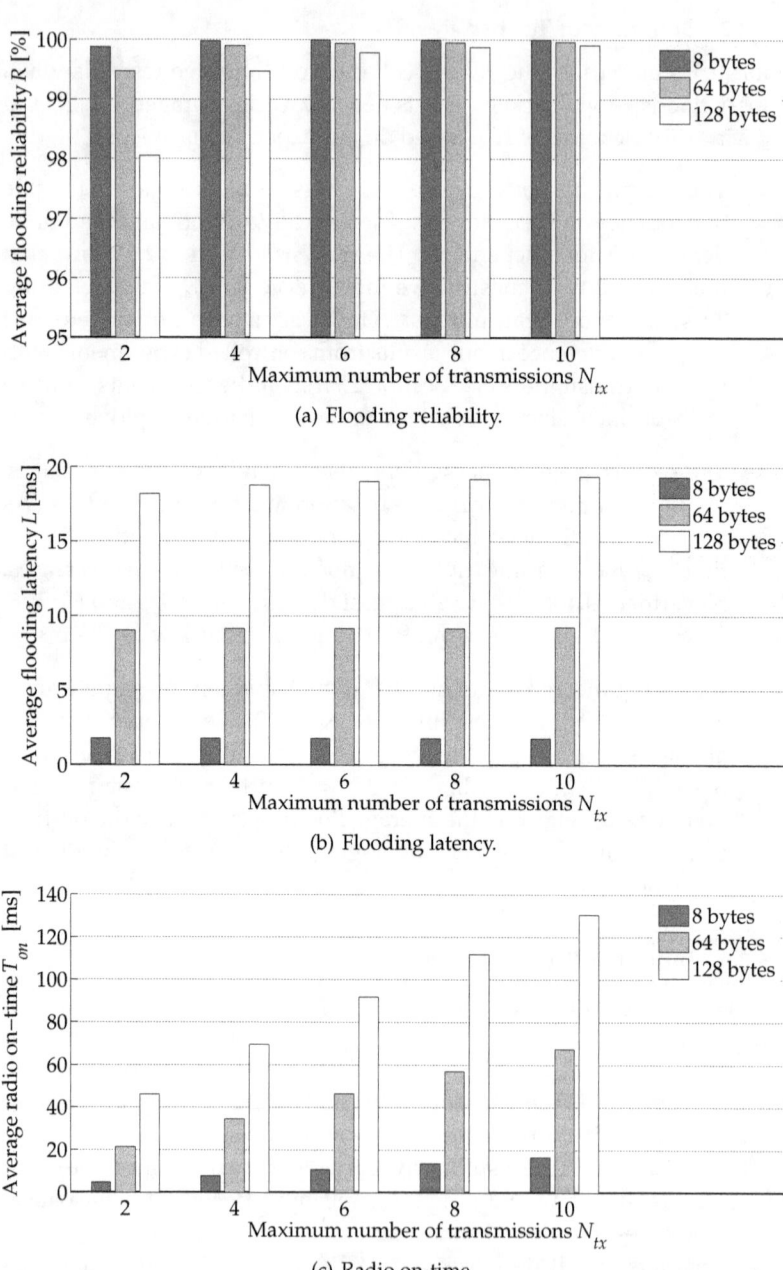

(a) Flooding reliability.

(b) Flooding latency.

(c) Radio on-time.

Figure 2.17:   Average Glossy performance on MoteLab for various packet lengths, with a transmit power of 0 dBm.

on-time increase linearly with packet length, as discussed in Section 2.3.2. This is because each single packet transmission and reception takes longer. Nevertheless, for $N_{tx} = 4$, Glossy requires an average flooding latency of only 18.77 ms and an average radio on-time of 69.59 ms to flood 128-byte packets. These results indicate that Glossy is suitable also for applications that need to flood long packets.

In the remaining of the chapter, we report results from TWIST and DSN based on network floods of 8-byte packets. This packet length is sufficient to send short commands to the nodes (e.g., to set system parameters).

2.6.2.4 Impact of Network Characteristics

We now look at the performance of Glossy under different network characteristics. To this end, we run experiments with three different transmit powers on TWIST and DSN. In doing so, we vary the average node density in the network and the network diameter. The maximum number of transmissions during a flood is fixed to $N_{tx} = 3$ across all runs.

Figure 2.18 plots the CDFs of R, L, and T_{on} on TWIST, for transmit powers 0 dBm, -15 dBm, and -25 dBm. Looking at Figure 2.18(a), we find that all receivers have a flooding reliability above 99.99 % at the highest power setting (i.e., at the highest node density). At the lowest power, 80 % of the receivers experience such high reliability. The drop in R is due to an increased network diameter. In fact, it takes 5 hops instead of 3 to reach the farthest receivers at the lowest power. This is also reflected in the step-wise shape of the CDF: each step corresponds to the flooding reliability experienced by receivers at a certain hop distance from the initiator. The results confirm also that R exhibits no noticeable dependence on node density, as hinted by controlled experiments in Section 2.6.1.2.

We see from Figure 2.18(b) that Glossy needs less than 3 ms to flood a packet to all 91 receivers, even at the lowest power that merely keeps the network connected. We are not aware of any current protocol that provides such fast flooding. We comment on related flooding and dissemination protocols in Section 2.7.

Finally, Figure 2.18(c) plots the radio on-time. We see that receivers listen longer as their hop distance from the initiator increases. Nevertheless, Glossy achieves ultra-low duty cycles also for larger network diameters. For example, consider an application that wants to use Glossy to (potentially) flood a command every minute. Then, on TWIST, Glossy would utilize not more than 0.01 % of a node's average radio duty cycle. We measure comparably low duty cycles on MOTELAB with a maximum distance of 8 hops from the initiator.

Figure 2.19 shows that Glossy achieves similar performance also on DSN, for transmit powers 0 dBm, -10 dBm, and -15 dBm. We see in

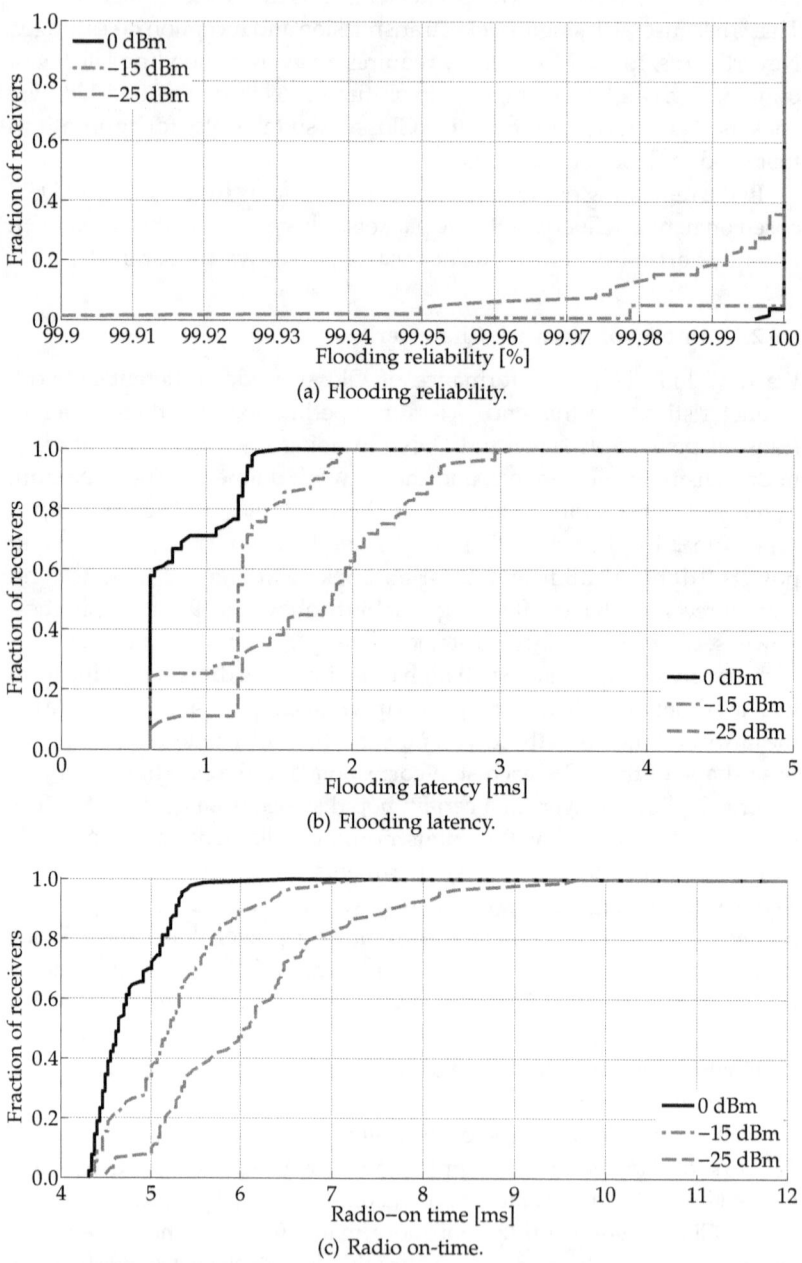

(a) Flooding reliability.

(b) Flooding latency.

(c) Radio on-time.

Figure 2.18: CDF of Glossy performance on TWIST with three different transmit powers, for $N_{tx} = 3$.

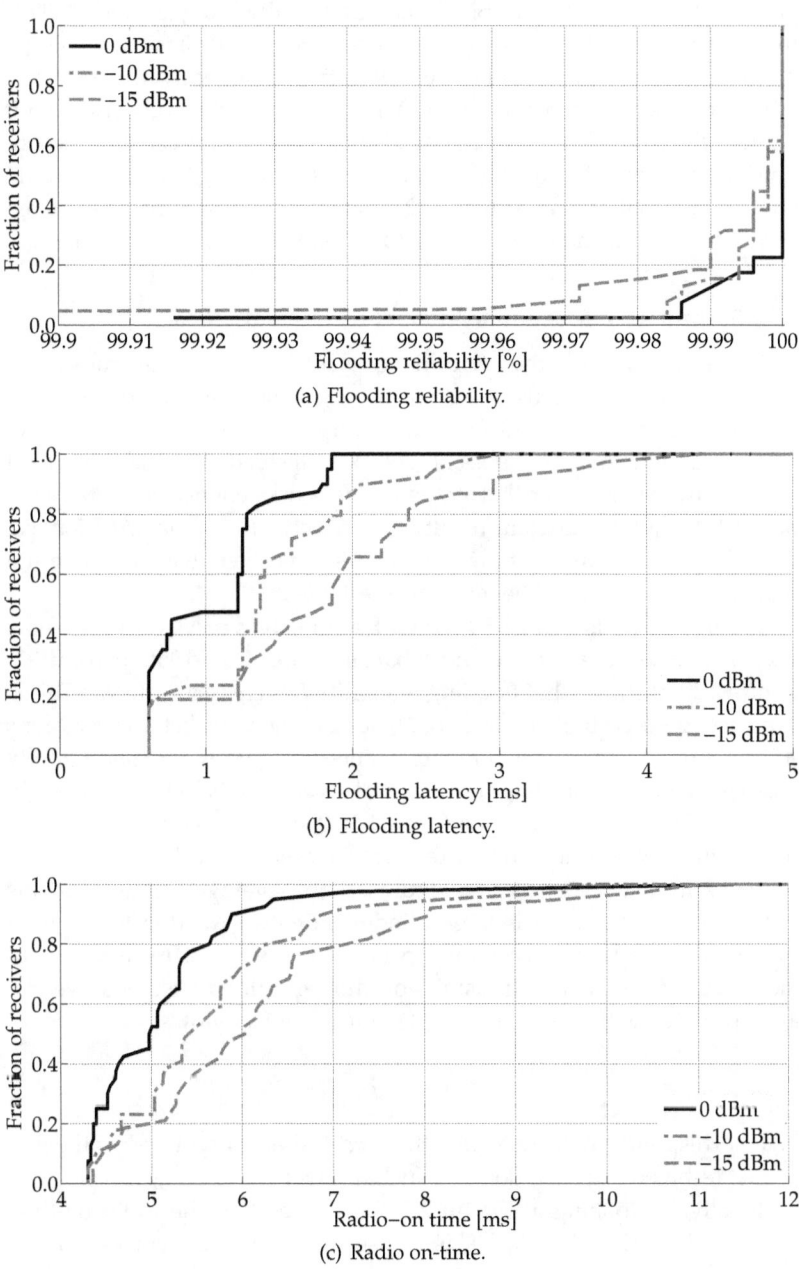

(a) Flooding reliability.

(b) Flooding latency.

(c) Radio on-time.

Figure 2.19: CDF of Glossy performance on DSN with three different transmit powers, for $N_{tx} = 3$.

Figure 2.19(a) that also on DSN flooding reliability is bigger than 99.99 % for at least 80 % of the receivers. Due to the lower node density and bigger diameter of DSN, we see in Figure 2.19(b) that more receivers experience a higher flooding latency than on TWIST already at the highest transmit power of 0 dBm. This corresponds also to a slight increase in radio on-time, as shown in Figure 2.19(c). Nevertheless, flooding latency and radio on-time are smaller than 5 ms and 12 ms at all receivers, respectively, even at the lowest transmit power of -15 dBm that keeps DSN connected.

2.6.2.5   Impact of Maximum Number of Transmissions

Next, we analyze how the performance is affected by N_{tx}, the maximum number of transmissions per node during a network flood. We run experiments on TWIST and DSN, and vary N_{tx} between 1 and 10. To stress Glossy as much as possible, on both testbeds we use the lowest possible transmit power that keeps the networks connected. On TWIST, nodes transmit at -25 dBm, resulting in a network diameter of 5 hops; on DSN, with a transmit power of -15 dBm, receivers have a maximum distance of 7 hops from the initiator (see Table 2.1).

Figure 2.20 plots R, L, and T_{on} for different values of N_{tx} on TWIST. Bars show network-wide averages; error bars indicate 5th and 95th percentiles. Figure 2.20(a) shows that flooding reliability R increases with N_{tx}. This is expected, because higher values of N_{tx} lead to more packet transmissions during a flood and thus to a higher chance for a receiver to successfully receive the packet. Starting from $N_{tx} = 3$, average reliability consistently exceeds 99.99 %. In fact, we performed 50,000 floods with $N_{tx} = 10$, and only in three cases one of the 91 receivers missed the packet.

Flooding latency, shown in Figure 2.20(b), averages around 1.75 ms for all values of N_{tx}. It is largely independent of N_{tx} due to the high probability of correctly receiving a packet already at the first attempt. In fact, assuming that each transmission during a flood succeeds, we can approximate the flooding latency L of a receiver as follows:

$$L \approx (c^* + 1) \times T_{relay} \tag{2.9}$$

T_{relay} corresponds to the relay length and c^* is the relay counter of the first packet received during a flood by such receiver.

Finally, by looking at Figure 2.20(c), we see that the radio on-time increases linearly with N_{tx}. This is also reflected in the corresponding analytical expression, which again assumes successful transmissions during a flood:

$$T_{on} \approx (c^* + 2 \times N_{tx}) \times T_{relay} \tag{2.10}$$

The term $c^* \times T_{relay}$ accounts for the time a receiver listens before the first transmission it overhears, and is independent of N_{tx}. The term $2 \times N_{tx} \times T_{relay}$ corresponds instead to the time a receiver spends in receiving and transmitting packets, and it increases linearly with N_{tx}. Even when setting the maximum number of transmissions to $N_{tx} = 10$, we measure an average radio on-time smaller than 17 ms.

Figure 2.21 shows that on DSN the performance of Glossy experiences similar dependence on the maximum number of transmissions N_{tx}. We see in Figure 2.21(a) that flooding reliability R increases with N_{tx} and is bigger than 99.99 % when N_{tx} is at least 4. Figure 2.21(b) shows that flooding latency is again mostly independent of N_{tx}, as suggested by the approximation in (2.9). Finally, from Figure 2.21(c) we see that the average radio on-time increases linearly with N_{tx}, in accordance with (2.10).

While the average flooding latency and radio on-time are very similar on TWIST and DSN, we see a more significant difference in their 95th percentiles. This is because some nodes on DSN are 7 hops away from the initiator, whereas on TWIST the network diameter is at most 5 hops. This leads to higher values of c^* for these nodes, and thus to higher values of L and T_{on} in (2.9) and (2.10).

2.7   Related Work

Using Glossy, nodes transmit the same packet synchronously. This idea stems from work on cooperative communication schemes [SMSM06]. However, requirements such as precise time synchronization among multiple transmitters have long been considered too demanding for an implementation on real sensor nodes [SZHT07].

Flury and Wattenhofer demonstrate the feasibility of signaling a binary value to all nodes with an unmodulated wave [FW10]. Constructive interference provides the opportunity to extend this to real data packets. Dutta et al. propose Backcast as an acknowledged anycast service [DMEST08]. Backcast exploits constructive interference of short acknowledgment packets automatically generated by the radio hardware. It does not require synchronization among the nodes, but the application has very limited control over the content of the interfering packets. Backcast serves as the basis for A-MAC, a receiver-initiated link layer protocol [DDHC+10]. Moreover, interference has been exploited to increase the throughput of wireless networks (e.g., through analog network coding [KGK07]).

Flash [LW09] uses concurrent transmissions for rapid flooding in sensor networks. Flash relies exclusively on capture effects, which

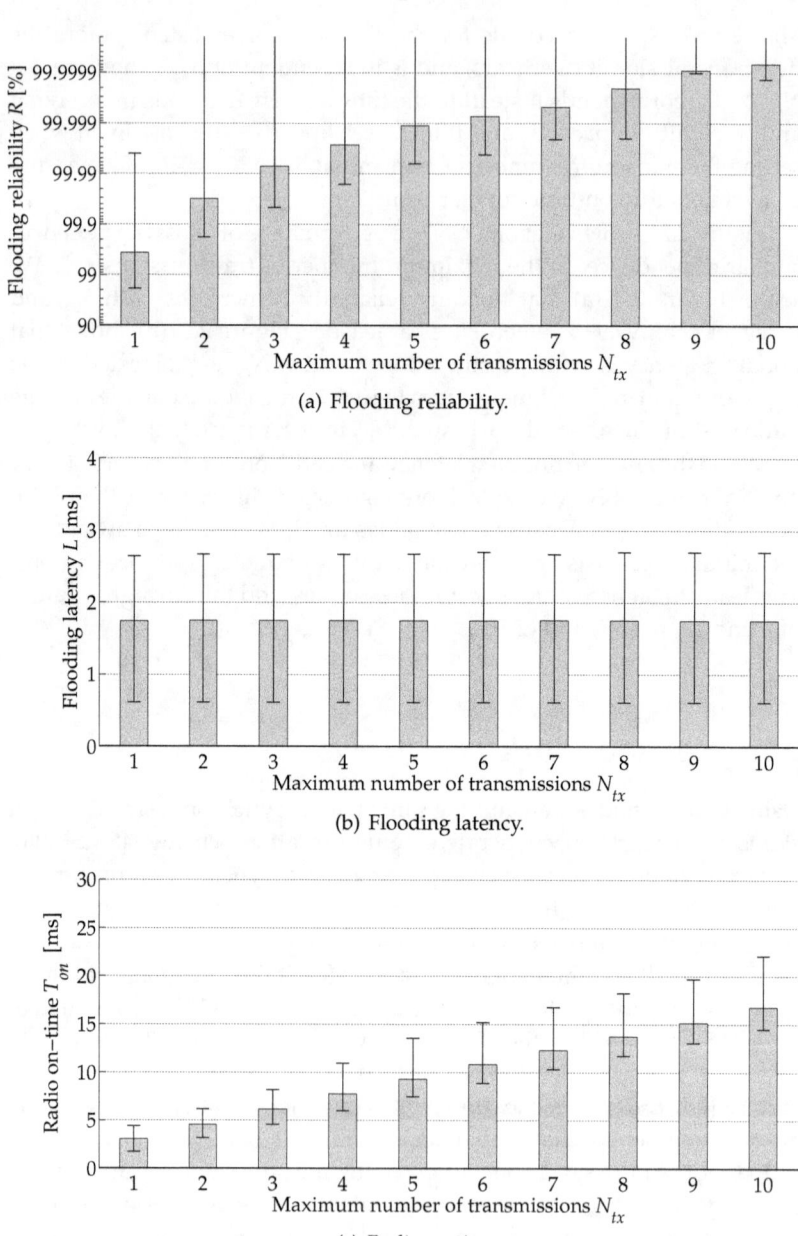

(a) Flooding reliability.

(b) Flooding latency.

(c) Radio on-time.

Figure 2.20: Glossy performance on TWIST for various values of N_{tx}, with a transmit power of -25 dBm. Bars denote averages; error bars indicate 5th and 95th percentiles.

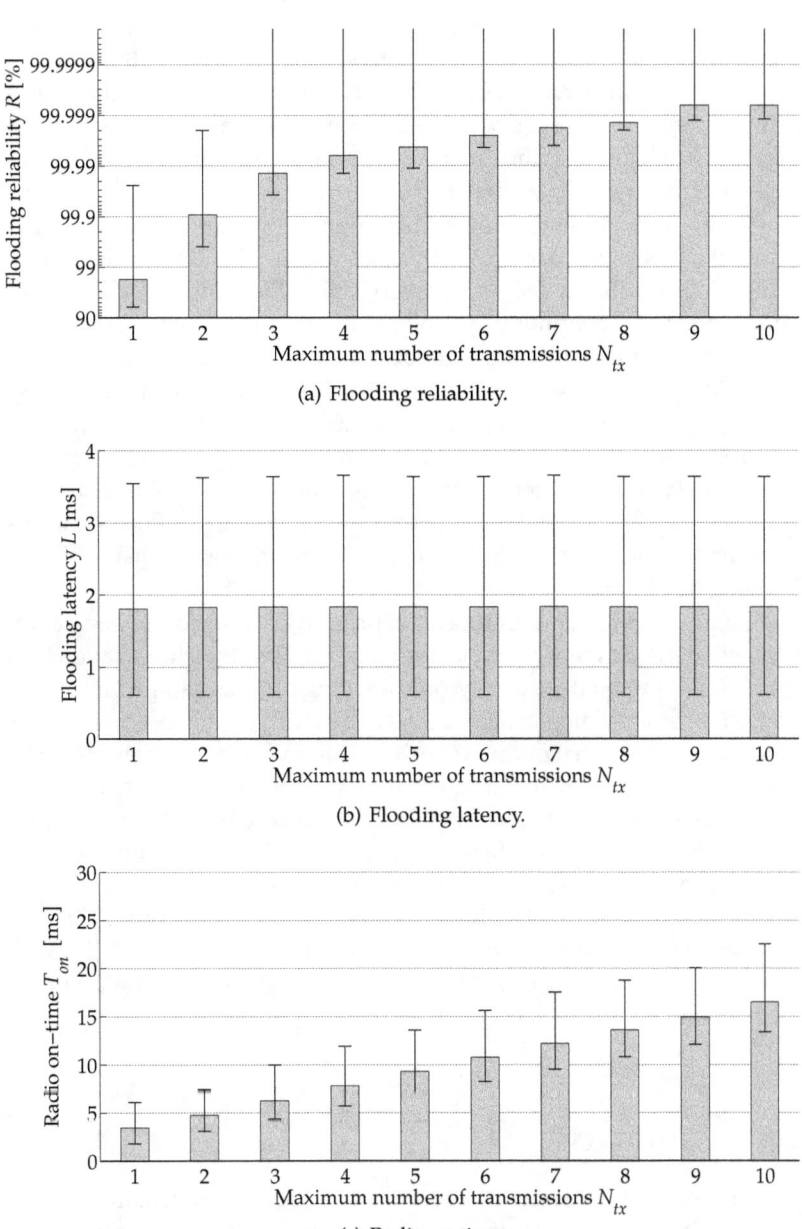

(a) Flooding reliability.

(b) Flooding latency.

(c) Radio on-time.

Figure 2.21: Glossy performance on DSN for various values of N_{tx}, with a transmit power of -15 dBm. Bars denote averages; error bars indicate 5th and 95th percentiles.

considerably reduces the chances of correct packet reception when many nodes transmit concurrently and forces Flash to control the number of transmitters [LW09]. Glossy also benefits from capture effects but primarily exploits constructive interference. This enables Glossy to flood packets with high reliability at any node density, as demonstrated by our testbed experiments in Section 2.6.2.

Glossy and Flash do not require nodes to maintain information about the network topology. By contrast, in the Robust Broadcast Protocol (RBP) [SHSM06] and the Collective Flooding (CF) [ZZHZ10] nodes need to continuously collect information about their local neighborhood to identify links important for the broadcast propagation.

Trickle [LPCS04] and its variants like Deluge [HC04], Drip [TC05], and DIP [LL08] provide data dissemination: nodes continuously send advertisements to detect new data and ensure complete network coverage. Typically, dissemination protocols are optimized for reliability and data consistency, not for latency or energy. Glossy floods packets fast without additional control traffic, while sacrificing less than 0.01 % in flooding reliability.

Flooding is a basic communication primitive for time synchronization in sensor networks. For example, the Flooding Time Synchronization Protocol (FTSP) [MKSL04] uses periodic flooding of time-stamped messages and achieves a per-hop synchronization error in the microsecond range. Lenzen et al. show that optimal synchronization necessitates fast network flooding [LSW09]. Their PulseSync protocol achieves a higher accuracy than FTSP and a flooding latency below one second. Glossy provides even higher accuracy by flooding packets within a few milliseconds and employing the Virtual High-resolution Time (VHT) approach by Schmid et al. [SDS10]. The high accuracy and low energy of VHT are also due to the use of a custom external high-speed crystal [SDS10]. Glossy could enable further improvements in synchronization accuracy by combining it with such crystals.

2.8   Summary

This chapter has proposed Glossy, a novel flooding architecture for low-power wireless networks that uses interference to its advantage. By making the baseband signal of synchronous transmissions of the same packet interfere constructively, Glossy enables receivers to decode a packet even in the absence of capture effects. We have analyzed the robustness of our techniques in achieving constructive interference based on a mixture of stochastic and worst-case models. We have evaluated our implementation of Glossy using experiments under controlled settings and on three wireless sensor testbeds. The results demonstrate that Glossy provides accurate time synchronization along with fast and highly reliable flooding at ultra-low duty cycles, showing no noticeable dependence on node density in the scenarios considered.

We have made the source code of Glossy publicly available at http://www.tik.ee.ethz.ch/~ferrarif/sw/glossy and, as part of the Contiki Projects Community, at http://sourceforge.net/p/contikiprojects/code/HEAD/tree/ethz.ch/glossy. Since its release, several independent researchers have exploited, extended, improved, or ported Glossy to other platforms, demonstrating its high impact on the sensor network community.

The unprecedented performance and timing accuracy of Glossy has fostered the design of innovative communication protocols for low-power wireless networks. TriggerCast, for example, is a flooding architecture that improve on Glossy by compensating for variable propagation and radio processing delays [WHC+13]. After porting Glossy to the CC430 platform (see Section 2.4.5), Carlson et al. incorporate a forwarder selection mechanism to let only a subset of nodes participate in a flood, thus improving energy efficiency and throughput in point-to-point communication [CCT+13]. In the Splash dissemination protocol, Doddavenkatappa et al. add channel diversity to Glossy and combine it with pipelining to disseminate large amount of data while being 20 times faster than state-of-the-art dissemination protocols [DCL13]. Using the hardware implementation of Glossy for the μSDR platform mentioned in Section 2.4.5, Kuo et al. are currently developing Floodcasting, a protocol envisioned to improve the range and coverage of visual light communication (VLC) networks and to achieve buffer-free, real-time audio streaming in multi-hop wireless networks [KPD13]. Finally, Landsiedel et al. use the Glossy source code as the basis for developing CAOS, a primitive that exploits capture effects and constructive interference to provide ultra-fast all-to-all communication and in-network processing in low-power wireless networks [LFZ13a].

The original Glossy architecture has also been used as a service or as the underlying primitive for more general communication protocols. In pTunes, for example, Zimmerling et al. exploit the extremely low latency and high energy efficiency of Glossy to periodically collect information about the current network state and disseminate optimized MAC parameters [ZFM+12]. In the next chapters of this thesis we show that Glossy is the one-to-all communication primitive we leverage to achieve a wireless bus that enables dependable communication in cyber-physical systems.

3

Low-Power Wireless Bus (LWB):
A Versatile Wireless Bus

Nodes in a cyber-physical system need to continuously interact and coordinate across multi-hop wireless networks in order to provide the desired level of dependability. These interactions are required, for example, to detect node and communication failures and agree on appropriate countermeasures. A wireless bus able to efficiently support multiple traffic patterns such as one-to-many, many-to-one, and many-to-many is thus paramount.

Emerging applications for low-power wireless may also benefit from a protocol that supports multiple communication patterns. These networks are indeed gaining momentum beyond early data collection applications. For instance, recent deployments demonstrate the feasibility of closed-loop control [CCD+11] and increasingly employ mixed installations of static and mobile devices [CLBR10, DEM+10]. These applications are characterized by a blend of traffic patterns, such as many-to-many communication for collecting sensor data at multiple receivers and one-to-many communication for disseminating control commands [CCD+11]. They also often feature end-to-end interactions across static and mobile nodes [CLBR10, DEM+10]. Looking at [CLBR10], for example, we observe the need to couple two communication protocols: one that delivers patient data from mobile sender nodes to static infrastructure nodes, and another one that forwards the data over the static infrastructure nodes towards a common receiver (i.e., the application receiver).

In contrast to the diverse needs of emerging applications, current communication protocols support specific traffic patterns (e.g., one-to-many [LPCS04], many-to-one [GFJ+09] or many-to-many [MP11]) in distinct scenarios (e.g., static networks [GFJ+09] or with receiver mobility [MSKG10]). This forces designers to form ad-hoc protocol ensembles to satisfy the application demands, which may entail adapting existing implementations [CCD+11] or developing custom protocols in absence of suitable off-the-shelf solutions [CLBR10]. As a result, multiple protocols that were designed in isolation need to operate concurrently. This is often detrimental to system performance [CKJL09], and causes protocol interactions that are difficult to cope with [WCL+07].

To address this problem, this chapter presents the *Low-Power Wireless Bus (LWB)*, a simple yet efficient communication protocol that provides a unified solution for several traffic patterns, while supporting also mobile nodes immersed in static infrastructures. LWB's design revolves around three cornerstones:

1. We exclusively use fast *network floods* for communication. This effectively turns a multi-hop low-power wireless network into a network infrastructure similar to a shared bus.

2. Similar to fieldbus-based communication protocols [KG93], we adopt a *time-triggered operation* to arbitrate access to the shared bus. Nodes are time-synchronized and access the bus according to a *global communication schedule* computed online based on current traffic demands.

3. We compute the communication schedule *centrally* at a dedicated *host* node. The host periodically distributes the schedule to all nodes to coordinate the bus operation.

To support our design, we use Glossy as the underlying flooding mechanism. As discussed in Chapter 2, Glossy provides high flooding reliability with minimal latencies, offering a foundation for 1. and 3., and accurate global time synchronization, which we leverage as a stepping stone for 2. Glossy also maintains no topology-dependent network state, which spares state reconfigurations when the network topology changes.

As a result, LWB simplifies the networking architecture by replacing the standard network stack with a single-layer solution that:

- *Supports multiple traffic patterns.* The exclusive use of Glossy network floods makes all nodes in the network potential receivers of all data. LWB leverages this opportunity to support both many-to-one and many-to-many traffic, besides the one-to-many pattern provided by Glossy itself. This occurs without changes to the protocol logic, and straightforwardly enables scenarios where, for example, multiple receivers are opportunistically deployed [VVV07].

- *Adapts to varying traffic demands.* The centralized, Glossy-based operation makes LWB efficiently support applications that adjust data rates at runtime, for example, in response to external stimuli [ADB+04]. Nodes indeed inform the host of changes in their traffic demands, and the host adapts at runtime the main protocol parameters and how bandwidth is distributed among nodes.

- *Is resilient to topology changes.* Different from most existing solutions, the network state kept at a LWB node is independent of the network topology and thus resilient to any such change. No state reconfigurations are indeed required to keep up with changing topologies, which reduces LWB's control overhead to a minimum. This provides efficient support to deal with link fluctuations, most notably due to node failures [BGH+09] and interference [LPLT10].

- *Supports node mobility.* As an extreme form of topology change, LWB encompasses also mobile nodes, acting as senders, receivers, or both, without any changes to the protocol logic. This applies to scenarios where, for example, mobile nodes interact with a fixed infrastructure [CLBR10].

The LWB protocol, described throughout Sections 3.1 to 3.3, renders our design concrete and complements it with mechanisms to: (*i*) ensure a fair allocation of bandwidth across all traffic demands; (*ii*) support nodes dynamically joining and leaving the system (e.g., due to node failures or disconnections); and (*iii*) resume communication after a host failure, thus overcoming single point of failure problems.

Using the same prototype, we evaluate in Sections 3.4 to 3.9 LWB's performance on four testbeds that range from 26 to 260 nodes, including a testbed with nodes attached to robots for repeatable mobility experiments. For comparison, we consider seven combinations of state-of-the-art routing and link-layer protocols: the Backpressure Collection Protocol (BCP) [MSKG10] over a non duty-cycled Carrier Sense Multiple Access (CSMA) layer; the Collection Tree Protocol (CTP) [GFJ+09] over CSMA, the Low-Power Listening (LPL) layer [PHC04], and A-MAC [DDHC+10]; Muster [MP11] over CSMA and LPL; and Dozer [BvRW07]. Based on 256 independent runs over a total duration of 838 hours, we find that:

- In many-to-one scenarios, LWB performs comparably to Dozer under light traffic and outperforms CTP in data yield and radio duty cycle; for example, LWB sustains traffic demands one message every 5 seconds from 259 senders with almost 100 % data yield, a situation where CTP + LPL collapses.

- In the same scenarios, LWB adapts promptly and efficiently to varying traffic demands; for example, when the aggregate traffic load suddenly increases from 54 to 460 messages per minute, LWB keeps data yield close to 100 %, whereas CTP + LPL and Dozer are significantly affected.

- In many-to-many scenarios, LWB outperforms Muster regardless of the number of senders or receivers, providing higher data yield than Muster + CSMA at a fraction of the radio duty cycle of Muster + LPL at all wake-up intervals.

- Under external interference and multiple concurrent node failures, LWB's performance is only marginally affected, whereas CTP and Dozer require routing state reconfigurations that cause significant performance loss.

- In the presence of mobile nodes, LWB outperforms BCP and CTP at no additional energy costs, delivering more than 99 % of the messages at very low radio duty cycles regardless of whether senders, receivers, or both are roaming.

Our results demonstrate that LWB is more versatile than existing communication protocols, and performs comparably or significantly better than the state of the art in all scenarios we tested. As such, LWB is directly applicable to a broad spectrum of low-power wireless applications, from data collection [CMP+09, LLL+09] to control [CCD+11] and mobile scenarios [CLBR10, DEM+10]. Most importantly, it serves as a foundation for the design of a dependable wireless bus that provides delivery guarantees, as we demonstrate in Chapter 4.

Under specific operating conditions such as linear topologies that span several tens of hops [KPC+07] and applications with mostly aperiodic traffic [ADB+04] LWB's efficiency decreases and dedicated solutions may perform better. We discuss the limitations of LWB in Section 3.10 by illustrating its scaling properties and the dependence of a few protocol parameters on the network diameter. We also present alternative scheduling policies to satisfy different application requirements. We review related work in Section 3.11 and summarize the contribution of this chapter in Section 3.12.

3.1 Overview

The LWB protocol completely replaces the standard network stack, sitting between radio driver and application.

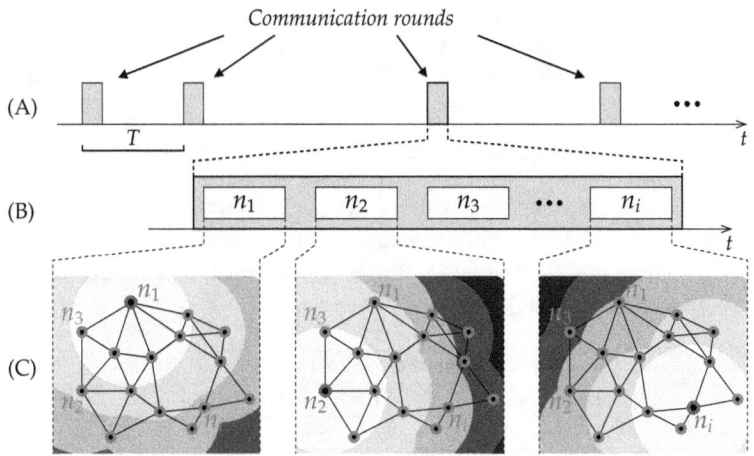

Figure 3.1: Time-triggered operation in LWB. Protocol operation is confined within communication rounds that repeat with a possibly varying round period T (A); each round consists of multiple non-overlapping slots (B); each slot corresponds to a distinct Glossy flood (C).

LWB in a nutshell. LWB maps all communication on Glossy floods. A single flood serves to send a message from one node to all other nodes. To avoid collisions among different floods, LWB adopts a time-triggered operation: nodes communicate according to a *global communication schedule* that determines when a node is allowed to initiate a flood.

LWB exploits Glossy's accurate global time synchronization. The protocol operation is confined within *communication rounds*. As shown in Figure 3.1 (A), rounds repeat with a possibly varying *round period T*, computed at the host based on the current traffic demands. Nodes keep their radios off between two rounds to save energy. Every round consists of a possibly varying number of non-overlapping *communication slots*, as shown in Figure 3.1 (B). In each slot, at most one node puts a message on the bus (initiates a flood), whereas all other nodes read the message from the bus (receive and relay the flood), as shown in Figure 3.1 (C). In LWB *every node participates in every flood*.

Figure 3.2 shows the communication slots within a generic round. A round r starts with a sched slot allocated to the host for distributing the communication schedule for that round. The schedule includes the round period and the mapping of individual nodes to the following data slots, if any. A req slot without a preassigned node follows; all sender nodes can contend in this slot, for example, to inform the host of their traffic demands. Based on the received traffic demands, the host computes the schedule for the next round $r + 1$—with a possibly updated round

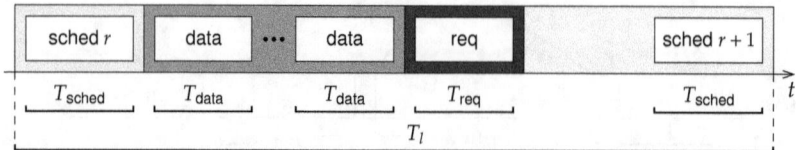

Figure 3.2: Communication slots within a generic LWB round r. Sender nodes use the req slot to inform the host of changes in traffic demands. After the req slot, the host computes the schedule for the next round $r + 1$, which it distributes already at the end of round r.

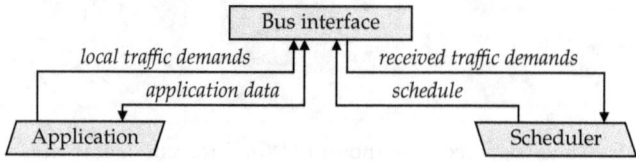

Figure 3.3: Conceptual architecture of a LWB node. The scheduler is present at all nodes but active only at the host.

period and mapping of sender nodes to data slots—and transmits the new schedule at the end of the round. The sched message for each round is thus transmitted twice, as we further illustrate in Section 3.2.4.

Application interface. The application interacts with LWB in two ways, as shown in Figure 3.3. First, LWB offers operations to place application messages in the outgoing queue for eventual transmission, and to receive incoming messages. Because of flooding-based communication, all nodes potentially receive all messages. At a sender node, the application specifies the intended recipients as a parameter to the send operation; at a receiver, LWB delivers a received message to the application only if the node is an intended recipient.

Second, LWB provides functions the application uses to notify LWB of changes in the traffic demands of a node. Targeting applications that feature mostly periodic traffic, LWB accepts traffic demands in the form of *periodic streams* of messages, defined by an *inter-message interval (IPI)* and a *starting time*. The application can dynamically change the traffic demands, creating new streams or stopping existing ones.

When a new traffic demand arises, the application issues a stream add request, specifying IPI and starting time. The latter may lie in the past if the application needs more data slots at the beginning, for example, to transmit a local backlog of messages. When a traffic demand ceases to exist, the application issues a stream remove request to cancel the stream. The application may issue multiple stream add requests from the same node (e.g., if different sensors produce readings at different rates) and may individually remove streams. LWB takes care of transmitting stream requests of either type to the host, where the scheduler component uses them as input for computing the communication schedule.

Next, Section 3.2 describes the LWB protocol operation, while Section 3.3 focuses on the scheduler, which is a stand-alone component present at all LWB nodes but active only at the host and responsible for determining the round period and computing the global communication schedule based on current application traffic demands.

3.2   Protocol Operation

To illustrate the protocol operation, we use sample executions of our LWB prototype whose implementation details are described in Section 3.4. We split the illustration according to the different phases an execution evolves into. These phases are purely for illustration purposes and do not correspond to distinct modes of protocol operation. Rather, the mechanisms we illustrate next blend together in a single protocol logic.

We start by illustrating in Section 3.2.1 the LWB operation in steady-state conditions, that is, when the host is informed of all traffic demands and these do not change over time. Next, we describe in Section 3.2.2 the bootstrapping phase that leads to such steady state. Finally, we describe in Section 3.2.3 how LWB adapts to reduce overhead should steady-state conditions endure for a given time. We discuss protocol optimizations in Section 3.2.4, mechanisms to handle communication and node failures in Section 3.2.5, and a strategy to overcome host failures in Section 3.2.6.

The scenario we consider for the non-failure case is a multi-hop network of 6 sender nodes and 1 receiver. All sender nodes have a stream with IPI = 6 s and starting time $t = 0$ s. The receiver acts also as the host.

3.2.1   Steady-State Conditions

Intuition. In steady state, nodes are time-synchronized and the host is aware of all traffic demands. We defer for the moment illustrating how the system reaches such state. The aggregated traffic demand in the scenario we consider amounts to 6 messages every 6 seconds. Say the round period T is 1 second and nodes are already informed of that. One way to schedule such traffic demands is to allocate 6 data slots, one per sender node, in one round every six. Other schedules are feasible, possibly with different round periods; here we simply illustrate a specific instance of our general scheduling strategy (see Section 3.3).

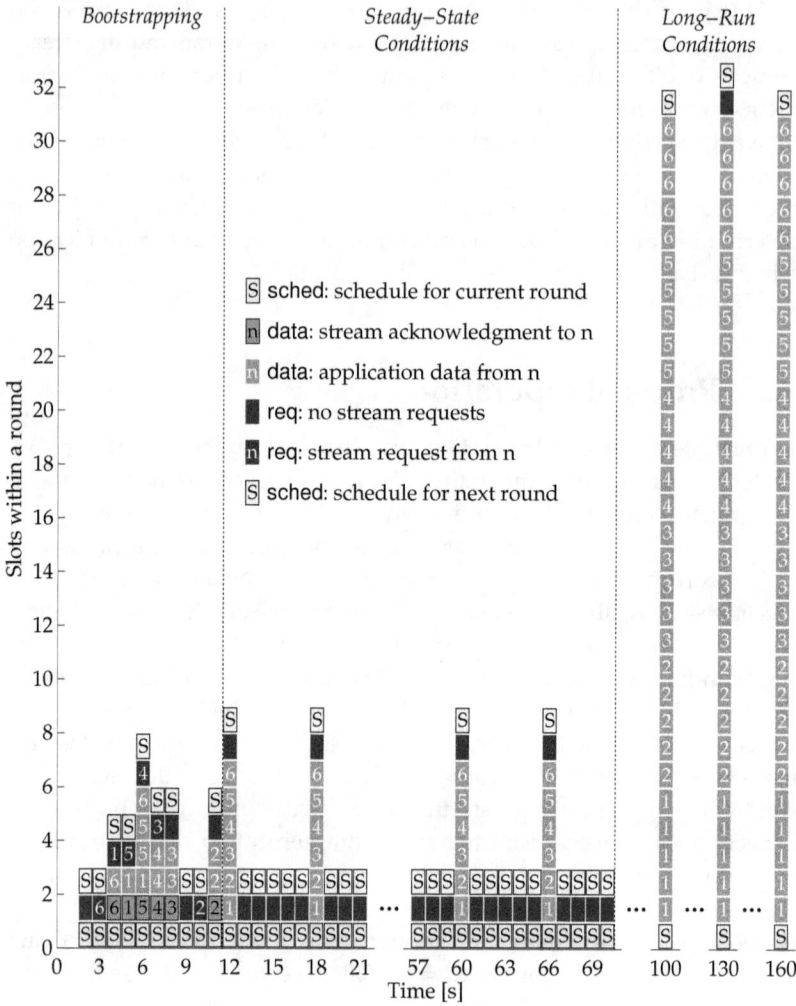

Figure 3.4: A trace of LWB's operation since startup when 6 sender nodes generate one data message every 6 seconds each. Nodes start with their radios turned on. Upon receiving a sched message for the first time at $t = 2$ s, nodes time-synchronize with the host and start duty cycling their radios. At $t = 11$ s, the host received all stream requests in the preceding req slots and starts allocating 6 data slots, one per sender, in one round every six. At $t = 70$ s, since no more stream request were recently received, the scheduler extends the round period T from 1 second to 30 seconds to reduce energy costs, and allocates 5 data slots to every sender in the following rounds; for the same reason, it allocates a req slot only every minute (i.e., every other round).

Steady traffic demands. The middle part of Figure 3.4 shows how the above materializes in a real LWB execution. Once steady-state conditions are reached at $t = 12$ s, the host distributes a schedule including 6 data slots, one per sender. Nodes periodically turn their radios on during communication rounds according to the round period $T = 1$ s. Based on the schedule, each sender accesses the bus during its allocated data slot and initiates a Glossy flood. All other nodes turn their radios on during every communication slot to relay the data messages. In addition, the receiver also delivers the messages to the application. These operations repeat every 6 seconds, as shown in the middle part of Figure 3.4.

The five rounds in between include no data slots. As described next, these seemingly redundant rounds are used to possibly receive further stream requests at the host. Should the host not receive new stream requests, rounds without data slots eventually disappear, as we illustrate in Section 3.2.3.

3.2.2   Bootstrapping

Intuition. To reach steady-state conditions, we need to: (*i*) time-synchronize all nodes with the host and inform them of the current round period T, and (*ii*) communicate the current traffic demands to the host. Nodes boot with their radios turned on and use the very first schedule transmission to synchronize initially. Afterward, sender nodes may use the req slot to communicate their traffic demands. Nodes can simultaneously access the bus during this slot, so communication may be unreliable when different Glossy floods overlap. We use a simple acknowledgement scheme to confirm that the host successfully received the traffic demands. As these are progressively received at the host, the scheduling of data slots intertwines with newly received stream requests.

Initial synchronization. At $t = 0$ s in Figure 3.4, only the host is part of the bus operation; all other nodes have their radios turned on. At $t = 2$ s, the first round starts with a sched slot where the host transmits a schedule for the first time. This sched message includes no mapping to data slots, but it only notifies about the presence of a req slot. Upon receiving the first schedule, Glossy time-synchronizes the nodes with the host, and nodes learn about the round period $T = 1$ s. This allows them to start duty cycling their radios and to effectively join the bus operation. Nodes that miss the first sched message keep their radios on until they finally receive the schedule in a subsequent round.

A pure relay node would stop the join operation at this point and from now on only help propagating the Glossy floods initiated by other nodes. Instead, a sender node continues the join operation, as described next.

Communicating traffic demands. During the early rounds in Figure 3.4, multiple sender nodes use the req slot concurrently to transmit their stream requests. In most cases, the host receives one stream request due to capture effects, which allow a node to receive a message despite concurrent transmissions [LF76]. This happens, for example, at $t = 3$ s in Figure 3.4, when the host receives a stream add request from node 6 during the req slot.

Based on this request, the host allocates two additional data slots in the next round at $t = 4$ s: one slot to itself to transmit a *stream acknowledgment* to node 6, and one slot to node 6 to transmit the application data it generated at $t = 0$ s. If no stream acknowledgments were received, node 6 would exponentially back off for some rounds before retransmitting the request. This reduces the number of contending requests in subsequent rounds, eventually increasing the chances of a successful reception by the host during a req slot.

Building up to steady state. At startup, the scheduler sets the round period T to the shortest possible to offer more req slots, speeding up the initial joining of sender nodes. Operations similar to the ones above repeat for node 1 at $t = 4$ s (add request during a req slot) and $t = 5$ s (stream acknowledgment and application data during data slots). At $t = 6$ s, the host allocates one data slot each to nodes 6 and 1 for transmitting their second application messages. It also allocates two data slots to node 5 in response to a stream add request received in the previous round: node 5 has two messages already generated (at $t = 0$ s, 6 s) to transmit. A similar processing occurs in the following rounds for nodes 4, 3, and 2.

Meanwhile, nodes are kept synchronized by the periodic transmission of sched messages. At $t = 11$ s, the host received all stream add requests and is thus aware of all traffic demands: the steady-state phase commences. Nevertheless, the host still schedules one req slot in each round for possible further requests. Indeed, the host does not know whether all senders have yet transmitted their stream requests. If new stream requests arrive later, the processing is the same described above.

3.2.3   Long-Run Conditions

Intuition. If steady-state conditions endure for a given time, the application has likely converged to a stable traffic pattern and load. This is the case in many scenarios we target [ADB+04, CCD+11, MCP+02, TPS+05], where periodic streams of data are initiated at startup and live on for the entire execution. This means that new stream requests are unlikely to arrive. In such situations, LWB minimizes control overhead by changing the schedule. Specifically, the host sets the round period T

such that: (*i*) LWB still provides enough bandwidth, and (*ii*) transmissions of sched messages occur sufficiently often to keep nodes synchronized.

Reducing overhead. At $t = 70$ s in Figure 3.4, the host detects that no new stream requests were recently received. The last stream request was indeed received at $t = 10$ s from node 2. It thus infers that the traffic demands are stable and increases the round period T from 1 second to 30 seconds. This value is based on the current traffic demands and the scheduling policy described in Section 3.3.

As a result, the following rounds occur every 30 seconds starting from $t = 100$ s. Increasing the round period reduces overhead, because it spares rounds with no data slots. At each round, the host indeed allocates $T/\text{IPI} = 5$ data slots to every sender node, corresponding to the number of data messages generated between consecutive rounds. LWB takes care of buffering these messages at the sender nodes until a data slot is available.

3.2.4   Optimizations

We complement the LWB operation just described with optimizations that further reduce the overhead and improve system responsiveness to changes in the traffic demands.

Transmissions of sched messages. Figure 3.4 already shows that sched messages are transmitted twice in a round. In principle, this is not necessary: the schedule transmitted at the beginning of a round would suffice to keep nodes synchronized and instruct them on when to access the bus. However, if the host changes the round period T as a result of computing the schedule for the next round, it would like to promptly communicate the new T to the nodes, so they can immediately switch to the new period (e.g., to save energy if the new round period is larger, as is the case at $t = 70$ s in Figure 3.4). We make this possible by letting the host transmit the sched message for the next round already at the end of the current round. Besides adding some redundancy in schedule transmissions, this improves responsiveness, because it makes nodes adapt earlier to updated round periods.

Scheduling req slots. Under stable traffic conditions similar to those in Section 3.2.3, new stream requests arrive rarely. Besides increasing the round period T, the host schedules req slots according to a different period T_r, whose value is an implementation constant independent of the current T. The right part of Figure 3.4 indeed already shows that req slots appear only once every $T_r = 1$ minute, at $t = 70$ s, 130 s, ... This optimization further reduces the overhead, especially when rounds unfold quickly to satisfy high, but stable, traffic demands.

Piggybacking stream requests. With long round periods, as in the right part of Figure 3.4, req slots occur rarely, and the optimization above makes them occur even more sparingly. Nevertheless, many nodes may compete in the req slot, and at most one at a time succeeds. These factors increase the latency in communicating changed traffic demands to the host. To ameliorate the problem, we let nodes piggyback stream requests on data messages if they are already assigned a data slot. This improves responsiveness, as it gives nodes more chances to send stream requests, and also reduces the pressure on req slots.

3.2.5   Node and Communication Failures

LWB needs to deal with node and communication failures, and nodes that spontaneously disconnect (e.g., due to mobility). Host failures, instead, require special countermeasures, which we discuss in Section 3.2.6.

Node failures and disconnections. If a node fails or disconnects from the network, its active streams are eventually reclaimed. LWB uses a simple counter-based scheme to detect such situations at the host. If the host does not receive any message within a certain number of consecutive rounds from a stream s, it removes s from the set of active streams. The threshold for removal is set as a protocol parameter. As a result, the scheduler stops allocating data slots to stream s, which saves bandwidth and energy.

This policy allows LWB to effectively detect situations where, for example, multiple nodes fail concurrently, as we show in Section 3.8.2. Due to Glossy's high reliability, our simple scheme is quite resilient to false positives: it is very unlikely that Glossy does not deliver data for a number of consecutive rounds while a node is still running and connected.

Communication failures. Our acknowledgment scheme ensures that all stream requests eventually reach the host. Problems may however arise if sched messages are lost. These are critical to keep nodes synchronized and to instruct them on when to access the bus.

To address this problem, LWB applies two policies. First, a node is allowed to participate in a communication round *only if* it received the schedule for the current round. Otherwise, it turns the radio off and keeps quiet for the remaining part of the round. Second, to compensate for a possibly higher synchronization error after a missed sched message, a node increases its guard times and wakes up slightly before the beginning of the next round. If a node misses the sched message for a given number of consecutive rounds, it continuously listens until it receives again a new schedule. In this situation chances are, for example, that the round period changed precisely in the missed schedule. The specific threshold for turning the radio on is a protocol parameter, and is set to 4 missed sched messages in our implementation. A detailed description of the synchronization state machine used in LWB is available in [ZFMT13].

Because of Glossy's high reliability, situations like those above happen very rarely, and usually indicate that a node is disconnected from the network or that the host has failed.

3.2.6   Host Failures

We address host failures by deploying the scheduler on *all* nodes, and by complementing LWB with mechanisms to dynamically enable or disable the scheduler at specific nodes according to a given policy. We describe next a simple failover policy that avoids multiple hosts being active on the same channel (e.g., when networks merge after partition).

Failover policy. Nodes detect a failure of the current host based on the *complete* absence of communication within a time interval T_{hf}, which is a protocol parameter that can be set by the user. If neither sched nor data messages are received within T_{hf}, it is very likely that also other nodes are not receiving schedules and thus that the host has failed.

We hardcode into all LWB nodes a circular ordered list of pairs $(channel, host\_id)$ that maps a set of communication channels to an appointed host for each channel. Upon detecting a host failure, a node switches to the channel next in the list. If the node is the appointed host for the channel, it activates the scheduler and starts distributing (empty) sched messages. Otherwise, the node turns the radio on and listens; if it receives sched messages, it joins the LWB operation on the new channel. In either case, if no communication is detected within another interval T_{hf}, the node switches to the next channel and the procedure repeats.

Our simple failover policy makes LWB remain functional despite repeated host failures. We note, however, that after a network partition several buses may operate on different channels and never merge again. More sophisticated failover policies, possibly based on self-stabilizing leader election [DIM97], can be developed to overcome this limitation.

Sample execution. We exemplify the functioning of our policy by inducing host failures in a real-world experiment. We use a multi-hop network of one receiver and 50 senders that generate messages with IPI = 1 minute. We set T_{hf} = 2 minutes, and use IEEE 802.15.4 channels 26, 15, and 25 with corresponding hosts H_1, H_2, and H_3. The circular ordered list stored at all nodes is thus: $\{(26, H_1), (15, H_2), (25, H_3)\}$. Initially, nodes use channel 26 and H_1 is the host.

Figure 3.5 shows the goodput at the receiver over time. Depending on failures and recoveries of the appointed hosts, the number of data messages received every round by the receiver varies as follows.

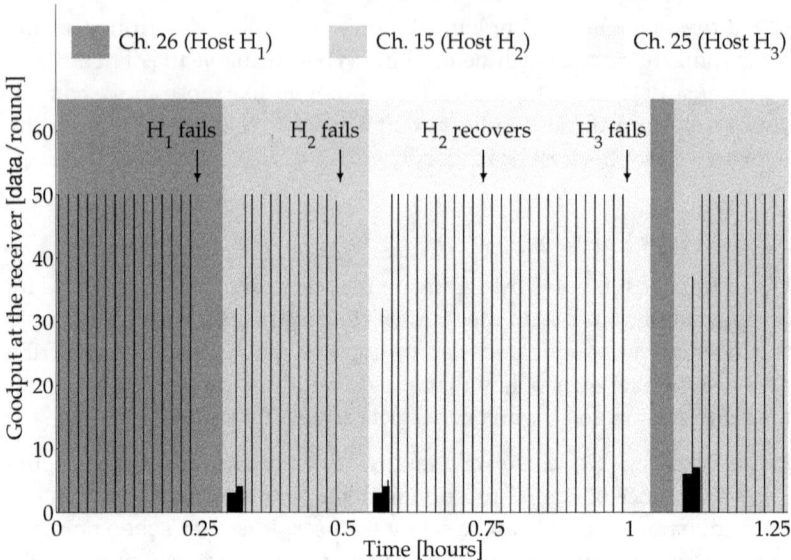

Figure 3.5: Goodput at the receiver as the number of data messages received per round when hosts fail and resume. A few minutes after detecting a host failure, communication resumes on a different channel with a new host.

- H_1 *fails at* $t = 0.25$ *hours*. Communication successfully resumes after $T_{hf} = 2$ minutes on the next channel in the list and with H_2 as the new host. All nodes join again, and senders issue stream add requests, considering also messages that were generated while they were disconnected from the bus. The new host eventually activates all streams and communication resumes as before the failure of H_1.

- H_2 *fails at* $t = 0.5$ *hours*. Similar events occur as after the previous host failure: after $T_{hf} = 2$ minutes with no communication, nodes switch to channel 25 and H_3 is the new host.

- H_2 *recovers at* $t = 0.75$ *hours*. This recovery has no visible impact on the bus operation. This is because nodes are operating on a different channel: after T_{hf} without receiving any stream request on channel 15, H_2 switches to channel 25 where it joins the ongoing LWB operation with H_3 as the host.

- H_3 *fails at* $t = 1$ *hours*. Communication resumes after $2 \times T_{hf} = 4$ minutes. Host H_1 for channel 26 has indeed not recovered yet, and a second timeout expires before nodes switch to channel 15 where H_2 is the new host.

3.3   Scheduler

The scheduler running at the host orchestrates communication over the bus by computing the communication schedule. This involves determining the round period T and allocating data slots to streams, as summarized also by the scheduler pseudocode shown in Figure 3.6. We describe next a scheduling policy that minimizes energy costs for low-power applications that can tolerate end-to-end latencies of a few seconds. We use this policy in Sections 3.4–3.9 to evaluate LWB. Section 3.10 presents alternative scheduling policies that trade smaller end-to-end latencies for slightly higher energy costs.

3.3.1   Determining the Round Period

Several trade-offs are involved in determining the round period T. It must be set sufficiently small to provide enough bandwidth for all traffic demands. However, the faster the rounds unfold, the higher is the energy overhead for distributing sched messages. We choose one specific design point: minimize the energy overhead under steady traffic conditions while satisfying all traffic demands whenever possible.

In addition, a LWB implementation on real devices imposes three constraints: (*i*) a lower bound T_{min} ensures that T is longer than the total *duration of a round* T_l (see Figure 3.2), the latter being an implementation-dependent constant; (*ii*) the round period T must also not exceed T_{max}, to ensure that nodes update their Glossy synchronization state sufficiently often; (*iii*) the number of data slots that the scheduler can map in a single sched message, and thus on the number of data slots it can allocate per round, is bounded by D, for example, due to platform-dependent restrictions on packet sizes.

Based on the above considerations, the scheduler computes the round period T as follows. To satisfy all traffic demands, it should allocate $R_{tot} = \sum_{s=1}^{S}(1/\text{IPI}_s)$ data slots per time unit, corresponding to the rate of data slots required by all S existing streams. To minimize the energy overhead for distributing sched messages, the scheduler should use the minimum number of rounds; that is, it should allocate *all* possible D data slots every round. The round period T_{opt} that minimizes energy while satisfying all traffic demands is thus:

$$T_{opt} = \frac{D}{R_{tot}} = \frac{D}{\sum_{s=1}^{S}(1/\text{IPI}_s)} \tag{3.1}$$

Shorter round periods can also satisfy all traffic demands, but entail more rounds and thus higher energy overhead. Longer round periods, instead, cannot satisfy all traffic demands.

variables

% current round
r: positive integer, initially 0;
% current number of active streams
S: positive integer, initially 0;
% current round period
T: positive integer, initially T_{min};
% data slots allocated during round r
K_r: set of data message identifiers;
% stream index
s: positive integer;
% number of data slots to allocate to the s-th stream
$toAlloc$: positive integer;

actions

% increment the round number
$r \leftarrow r + 1$;
% add/remove streams if add/remove stream requests were received;
% remove streams if sender failures were detected (see Section 3.2.5)
$S \leftarrow$ **updateSetOfStreams()**;

Determining the round period (Section 3.3.1)
% compute the round period that minimizes energy as per (3.1)
$T_{opt} \leftarrow$ **computeTopt**(S);
% check if the system is saturated
if $T_{opt} < T_{min}$: $sat \leftarrow$ **true**;   else: $sat \leftarrow$ **false**;
% update the current round period as per (3.2)
$T \leftarrow$ **updateT**(T_{min}, T_{max});

Allocating data slots to streams (Section 3.3.2)
% the schedule is initially empty
$K_r \leftarrow \emptyset$;
% draw a random stream index s between 1 and S
$s \leftarrow$ **random**(S);
% allocation of data slots, starting from the s-th active stream
for all S active streams:
    % compute the number of data slots not yet allocated to the s-th stream,
    % which corresponds to (3.5) in case of saturation
    $toAlloc \leftarrow$ **getSlotsToAllocate**(s, sat);
    % only $D - |K_r|$ data slots are still available
    $toAlloc \leftarrow$ **min**($toAlloc, D - |K_r|$);
    % add to the schedule the $toAlloc$ data slots allocated to the s-th stream
    $K_r \leftarrow K_r \cup$ **allocateDataSlots**($s, toAlloc$);

% possibly allocate a req slot, according to the policy in Section 3.2.4
allocateReqSlot()

Figure 3.6: Scheduler pseudocode. At each round, the scheduler running at the host computes the communication schedule for the next round by determining the round period T and allocating data slots to streams

Due to constraints (*i*) and (*ii*) above, the scheduler bounds T_{opt} in (3.1) within T_{min} and T_{max}. Finally, to limit the set of valid round period values, the scheduler sets T to the largest previous multiple of 1 second:

$$T = \left\lfloor \min\left(T_{max}, \max\left(T_{opt}, T_{min}\right)\right)\right\rfloor \tag{3.2}$$

If $T_{opt} < T_{min}$, the number of available data slots is insufficient for the current traffic demands: the network is *saturated*. When the scheduler detects saturation, besides setting the round period to $T = T_{min}$, it embeds this information into the sched message. This allows sender nodes to take appropriate actions if needed, such as temporarily storing data messages generated by the application in external memory to prevent queue overflows.

We complement the solution above with a simple policy to promptly react to varying traffic demands. If stream requests were recently received (e.g., in the last minute), the scheduler sets T to T_{min} regardless of the current traffic demands, and allocates a req slot in every round in anticipation of further stream requests. Such conditions occur, for example, when bootstrapping a network, as shown in Section 3.2.2, or when a subset of nodes wishes to send data messages at higher rates, as we show in Section 3.6.3.

3.3.2   Allocating data Slots to Streams

The scheduler allocates data slots to maximize fairness across all streams according to Jain's fairness index [JCH84], a metric widely used in the literature [RWAM05, WCB01]. Other metrics or algorithms can be applied by modifying the scheduler.

To compute Jain's fairness index, we denote with a_s the number of data slots the scheduler *allocates* to stream s during a round, and with $d_s = T/IPI_s$ the number of data slots stream s *demands* every round. The *allocation* to stream s is thus $x_s = \min(a_s/d_s, 1)$ [JCH84]. Using the allocations of all S streams, Jain's fairness index is defined as:

$$f(x) = \frac{\left(\sum_{s=1}^{S} x_s\right)^2}{S \times \sum_{s=1}^{S} x_s^2} \tag{3.3}$$

A fairness index of 1 indicates that the scheduler is equally fair to all streams; smaller values indicate less fairness.

As we illustrate next, our slot allocation scheme achieves a fairness index of 1 in the long run, as it is in general not possible to achieve fairness in individual rounds. For instance, a fair allocation in individual rounds may require allocating a non-integer number of data slots to streams.

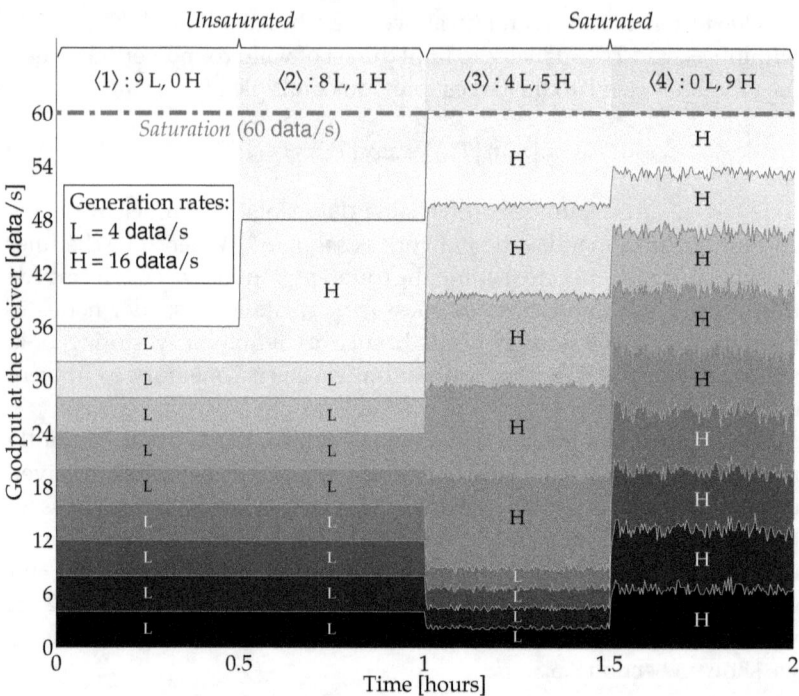

Figure 3.7: Goodput at the receiver as the number of data messages received per second when 9 sender nodes generate varying amount of traffic. LWB always allocates data slots fairly, based on traffic demands and available bandwidth.

In the following, we use an experiment on a multi-hop network of 10 nodes to exemplify how the scheduler allocates data slots in both the unsaturated and saturated case. One node acts as host and receiver. Each of the other 9 nodes generates one stream of 15-byte data messages, either at low rate L of 4 messages per second or at high rate H of 16 messages per second. These generation rates follow patterns of four phases $\langle 1 \rangle$–$\langle 4 \rangle$ in Figure 3.7, where different nodes generate data at different rates. In our LWB prototype, we set $T_{min} = 1\,s$, $T_{max} = 30\,s$, and $D = 60$ data slots.

Unsaturated network. When the network is not saturated, the scheduler can allocate sufficient data slots to satisfy all traffic demands, that is, $a_s = d_s$ for all streams s. This also implies that the allocation of data slots is eventually fair across all streams: $f(x) = 1$, as $x_s = 1$ for all streams s.

In the example experiment, the network is not saturated in phases $\langle 1 \rangle$ and $\langle 2 \rangle$. In phase $\langle 1 \rangle$, all 9 streams generate messages at low rate L: using (3.1), the scheduler computes $T_{opt}\langle 1 \rangle = 1.67\,s > T_{min}$. In phase $\langle 2 \rangle$, one node generates data at higher rate H, and $T_{opt}\langle 2 \rangle = 1.25\,s$ is still greater than T_{min}. The scheduler sets $T = T_{min} = 1\,s$ in both phases, and allocates data slots as follows.

- *Phase* $\langle 1 \rangle$. All streams demand 4 data slots every round and the scheduler satisfies this demand by allocating in total 36 data slots, 4 to each stream. Sender nodes indeed contribute equally to the total goodput at the receiver, as shown in Figure 3.7.

- *Phase* $\langle 2 \rangle$. One sender node increases its data rate to H. The scheduler allocates every round 16 data slots to it and 4 to the other nodes. In total, 48 data slots are allocated, and all traffic demands are satisfied.

Saturated network. If the network is saturated, the scheduler sets the round period T to the lower bound T_{min} but cannot satisfy all traffic demands. This means $a_s < d_s$ for at least one stream s. The scheduler allocates a_s data slots to each stream s such that the provided bandwidth is maximized and the allocation is eventually fair across all streams:

$$\text{allocate } a_s \text{ data slots such that } \sum_{s=1}^{S} a_s = D \text{ and } f(x) = 1 \qquad (3.4)$$

It can be shown that allocating the following number of data slots to each stream s is a solution to (3.4):

$$a_s = T_{opt}/\text{IPI}_s \qquad (3.5)$$

Intuitively, this entails allocating data slots to streams proportionally to their data rates. This allocation is fair because $x_s = a_s/d_s = T_{opt}/T_{min}$ is constant across all streams s, and hence $f(x) = 1$ according to (3.3).

The network in our example experiment is saturated during phases $\langle 3 \rangle$ and $\langle 4 \rangle$ in Figure 3.7, because the corresponding T_{opt} are smaller than T_{min}. The scheduler sets $T = T_{min}$ and allocates data slots to streams as follows.

- *Phase* $\langle 3 \rangle$. Four streams generate messages with rate H and four with rate L, leading to $T_{opt}\langle 3 \rangle = 0.625\,s$. Streams demand 96 slots per second altogether, but the scheduler can allocate at most $D = 60$ slots. To be fair, it allocates slots proportionally to their rates; that is, on average, $0.625 \times 16 = 10$ data slots to each of the 5 streams with rate H and $0.625 \times 4 = 2.5$ data slots to each of the 4 streams with rate L, against a demand of 16 slots and 4 slots, respectively.

- *Phase* $\langle 4 \rangle$. All streams generate data at rate H, and the network is saturated: $T_{opt}\langle 4 \rangle = 0.417\,s$. The scheduler allocates on average $a_s = 6.67$ data slots to every stream, against a demand of $d_s = 16$ slots. The sender nodes equally contribute to the goodput at the receiver, as shown in Figure 3.7.

Parameter	Description	Value
T_{min}	Minimum round period	1 s
T_{max}	Maximum round period	30 s
T_r	Period of req slots under stable traffic conditions	60 s
T_{sched}	Length of a sched slot	15 ms
T_{data}	Length of a data slot	10 ms
T_{req}	Length of a req slot	10 ms
N_{tx}	Maximum number of transmissions during a slot	3
D	Maximum number of data slots per round	60

Table 3.1: Default configuration of our LWB prototype.

Worth noticing is that the data slot allocation in the saturated case is fair because it applies the same allocation x_s to all streams s: $x_s = 10/16 = 2.5/4$ in phase $\langle 3 \rangle$ and $x_s = 6.67/16$ in phase $\langle 4 \rangle$.   Figure 3.7 indeed shows that during both these phases nodes with the same rate have the same goodput, and the four nodes with rate L in phase $\langle 3 \rangle$ have together the same goodput as one node with rate H, for L = H/4.

3.4   Evaluation Methodology

Before presenting experimental results, we describe the metrics, protocols, and testbeds we use to evaluate LWB.

Metrics.   We consider two key performance metrics commonly used for evaluating low-power wireless communication protocols [BvRW07, GFJ+09, MSKG10, MP11]:

- *Data yield,* defined as the fraction of application messages successfully received at the receiver(s) over those sent.

- *Radio duty cycle,* computed as the fraction of time a node keeps the radio on.

The former is an indication of the level of service provided to applications in delivering sensed data, whereas the latter provides a measure of a protocol's energy efficiency [AY05].

To determine data yield and radio duty cycle, we embed message sequence numbers and radio timings into data messages. We measure the radio duty cycle in software, using Contiki's power profiler and a similar approach in TinyOS. For each experimental setting and protocol, we compute these metrics based on three independent runs and report per-node or network-wide averages and 5th and 95th percentiles.

Protocol	Code footprint
LWB	22 kB
CTP + {CSMA, LPL, A-MAC}	{26, 28, 27} kB
Dozer	38 kB
Muster + {CSMA, LPL}	{35, 37} kB
BCP + CSMA	23 kB

Table 3.2: Code footprints of all protocol configurations used in the evaluation.

Protocols. We implement our LWB prototype on top of the Contiki operating system [Conb, DGV04], targeting the TelosB platform [PSC05]. We set the LWB configuration parameters as in Table 3.1. We discuss our choice for T_{sched}, T_{data}, and T_{req} in Section 3.10.2.

We compare our LWB prototype with seven combinations of routing and link-layer protocols, which represent the current state of the art.

- The Collection Tree Protocol (CTP) [GFJ+09] is a staple reference for many-to-one scenarios. We run CTP over a non duty-cycled Carrier Sense Multiple Access (CSMA) layer, the Low-Power Listening (LPL) [PHC04] layer, and A-MAC [DDHC+10]. CSMA serves as a baseline for CTP's data yield performance, since it provides the highest network capacity. The LPL setting matches the configuration used in [GFJ+09]. A-MAC is a receiver-initiated link layer shown to outperform LPL when running CTP [DDHC+10].

- Dozer is a TDMA-based collection protocol for periodic, low-rate many-to-one scenarios [BvRW07]. It achieves ultra-low radio duty cycles of 0.07–0.34 % in real deployments [KWL+11]. To compare Dozer with LWB, we port the original TinyNode implementation to the TelosB platform. Our results confirm that our port performs comparably to the original implementation under similar settings.

- Muster is one of the few protocols for many-to-many communication in low-power wireless networks tested on real nodes [MP11]. We run Muster atop LPL and CSMA. The LPL setting matches the configuration in [MP11], and the CSMA case serves as a baseline for Muster's data yield performance.

- The Backpressure Collection Protocol (BCP) represents the state of the art in data collection at a single mobile receiver [MSKG10]. Results indeed suggest that BCP outperforms recent mobile receiver routing protocols in terms of data yield [LKA+10, MSKG10, SGE06]. We run BCP atop CSMA, matching the configuration in [MSKG10].

Table 3.2 lists code footprints of all protocols, including test applications.

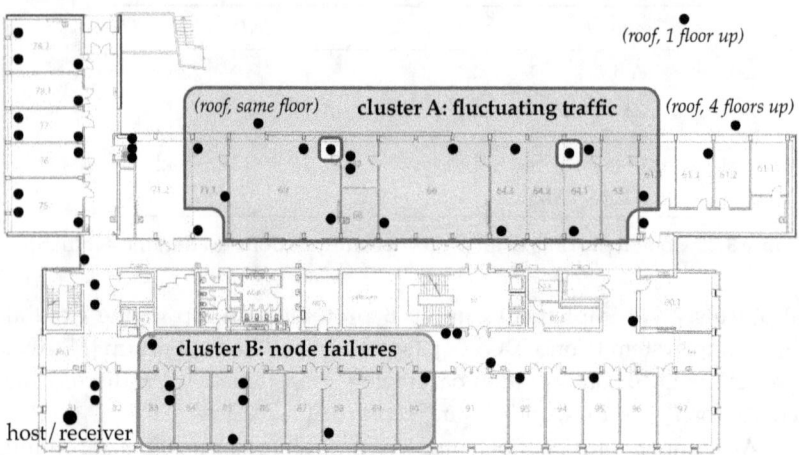

Figure 3.8: Layout of the FLOCKDSN testbed.

We use an application payload of 15 bytes in all experiments. The size of the control header is instead a function of the individual protocol. Unless otherwise stated, we neglect the bootstrapping phase, and start measuring after 10 minutes with LWB, after 0.5 hours with CTP and Muster, after 2 hours with Dozer, and after 15 minutes with BCP, giving each protocol enough time to discover the network and stabilize. We evaluate the bootstrapping performance separately in Section 3.5.

Testbeds. We use four sensor network testbeds: TWIST [HKWW06], KANSEI [EAR+06], the CONET integrated testbed (CONETIT) [CONa], and FLOCKDSN, a short-lived mixed installation of DSN [DBK+07] and FLOCKLAB [LFZ+13b]. As shown in Figure 3.8, FLOCKDSN consists of 52 indoor and 3 outdoor nodes. The figure also highlights the position of the node that serves as host and receiver, and of two clusters of nodes used in the experiments described in Section 3.6.3 and Section 3.8.2.

All testbeds feature TelosB nodes but differ along several dimensions as shown in Table 3.3, including number of nodes, node density, network diameter, and node mobility. KANSEI is the largest testbed we were able to gain access to. The 5 mobile nodes in CONETIT are attached to robots, allowing for repeatable mobility patterns.

The network diameter in Table 3.3 is based on the *physical* topology, matching LWB's perception; the maximum route stretch with other protocols is typically larger. Using a received signal strength indicator (RSSI) scanner, we find that on FLOCKDSN channel 20 is most exposed to Wi-Fi traffic. We use this channel in Section 3.8.1 to assess a protocol's vulnerability to external interference, whereas we use channel 26 in all other experiments to minimize Wi-Fi interference.

| Testbed | Location | Nodes | | Tx power | Diameter |
		Static	Mobile		
Twist	TU Berlin	90	0	-7 dBm	3 hops
Kansei	Ohio State University	260	0	-20 dBm	4 hops
ConetIT	University of Seville	21	5	-25 dBm	3 hops
FlockDSN	ETH Zurich	55	0	0 dBm	5 hops

Table 3.3: Testbeds used in the evaluation.

3.5 Bootstrapping

Bootstrapping is a critical phase in real deployments, because nodes may already spend a considerable amount of energy merely on commencing communication. By examining this facet of LWB, we find that:

Finding 1. *LWB bootstraps quickly and efficiently, while distributing energy costs equally among nodes.*

The following study also serves as a complement to the rest of the evaluation: starting from Section 3.6, we exclude the bootstrapping phase from the analysis not to bias the results.

Scenario. We consider a many-to-one scenario on Twist, where 89 senders start generating messages with IPI = 1 minute. Nodes log every second their current radio duty cycle into the local flash memory, and dump these logs over the serial port after 30 minutes. We run tests with LWB, Dozer, and CTP over A-MAC and LPL. We use the default 30 s beacon interval in Dozer [BvRW07], and set the wake-up intervals of A-MAC and LPL to 250 ms and 200 ms, which provide a good trade-off between data yield and energy consumption in this setting. As in all our experiments, we perform three trials with each protocol.

Results. We consider the systems fully bootstrapped when all 89 sender nodes delivered at least one message to the receiver. Figure 3.9(a) shows that LWB and CTP (independently of the link layer) bootstrap in roughly 2 minutes, whereas Dozer requires more than 18 minutes. Our results also indicate that LWB bootstraps most energy-efficiently: during the first 40 minutes of operation, nodes accumulate an average radio on-time of 34 s with LWB against 129 s with Dozer, 169 s with CTP + A-MAC, and 173 s with CTP + LPL.

Figure 3.9(b) shows a fine-grained analysis of energy costs by plotting the instantaneous radio duty cycle averaged across all nodes over the first 5 minutes of operation, which correspond to the grey region in Figure 3.9(a). The high energy efficiency of LWB is due to the scheduler *i)* setting the round period to $T_{min} = 1$ s on startup, which allows nodes to

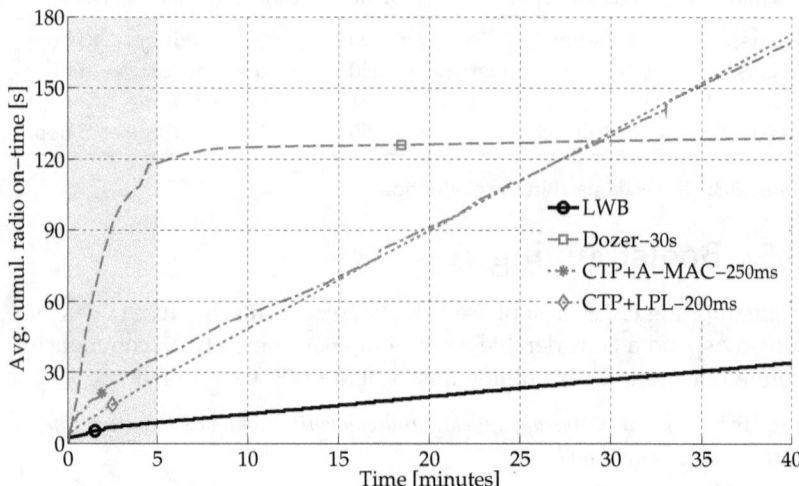

(a) Cumulative radio on-time, averaged across all nodes. Markers denote when the system has fully bootstrapped, that is, when all senders have delivered at least one application message to the receiver.

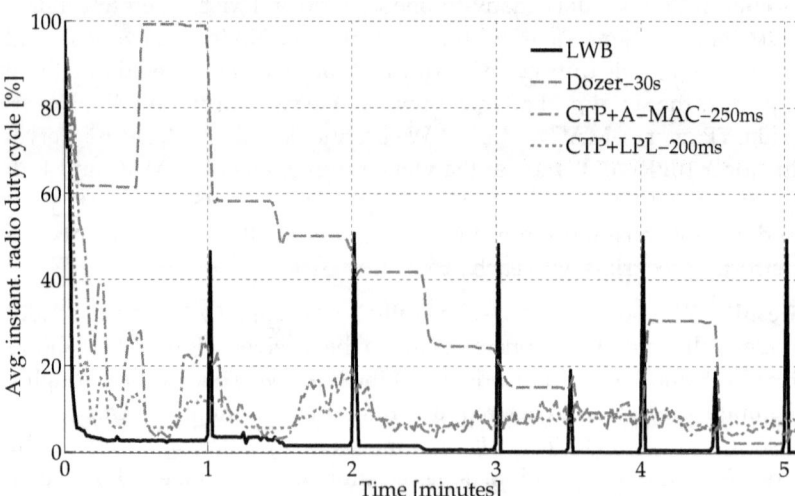

(b) Instantaneous radio duty cycle, averaged across all nodes. Peaks in LWB correspond to rounds where data slots are allocated to sender nodes.

Figure 3.9: Average performance during bootstrapping. LWB lets nodes join fast initially, and saves energy by quickly adapting the round period.

quickly time-synchronize and start duty cycling their radios after a few seconds; and *ii)* increasing the round period to $T = 30$ s and allocating fewer req slots when all stream add requests have been served, which further reduces energy costs. The initial synchronization is instead very expensive in Dozer, because of its fixed beaconing period of 30 s.

CTP's adaptive beaconing [GFJ$^+$09] ameliorates the problem, but still requires nodes to transmit broadcasts frequently during the first seconds. Broadcasts are costly over link layers like LPL and A-MAC, as visible from the peaks with increasing period in Figure 3.9(b). In Dozer, the synchronization between parent and children in the tree compounds the problem, because nodes need to keep the radio on for a full beacon period to discover their neighbors. This scanning phase is visible in the step-wise pattern in Figure 3.9(b), particularly between 30 s and 1 minute where the average radio duty cycle is 100 %.

Finally, we find that Dozer and CTP distribute the energy load unevenly among nodes, because they need to acquire information about the network topology and build a routing hierarchy. After 40 minutes, the difference between the maximum and minimum cumulative radio on-time of a node is 235 s with Dozer, 240 s with CTP + A-MAC, and 201 s with CTP + LPL. This may later cause a network partition due to faster battery depletion at nodes nearby the receiver [PGZM12]. Thanks to absence of a routing hierarchy, this difference with LWB is less than 27 s, which makes LWB largely immune to this problem.

3.6   Many-to-One Communication

In this section, we investigate the performance of LWB in many-to-one scenarios under varying traffic loads, which represent a significant fraction of existing low-power wireless applications [ADB$^+$04, LLL$^+$09, MCP$^+$02, TPS$^+$05]. Our results indicate that:

Finding 2. *In many-to-one scenarios, LWB operates reliably and efficiently under a wide range of traffic loads, and promptly adapts when the traffic demands change over time.*

A key aspect to understand the following performance results is that radio activity in LWB is exclusively driven by the global communication schedule. This spares nodes from periodically waking up merely for probing the channel, as in contention-based protocols like LPL or A-MAC. Placing this observation in perspective: with the energy budget A-MAC requires solely for probing the channel every 500 ms, LWB supports more than 50 streams with IPI = 1 minute (see Section 3.10.1.3).

3.6.1   Light Traffic

We first look at a common scenario for low-power wireless sensor networks: periodic, low-rate data collection at a single receiver (sink). This is typical of environmental monitoring, where high data yield and energy efficiency are paramount [MCP+02, TPS+05].

Scenario. We use FLOCKDSN and let 54 senders generate messages with IPI = 2 minutes for 4 hours. We compare LWB with Dozer and CTP over A-MAC and LPL. Given the light traffic load and stable network conditions in this scenario, we use Dozer's default 30 s beacon interval, and set the wake-up interval in A-MAC and LPL to 500 ms.

Results. Figure 3.10 plots the cumulative distribution functions (CDFs) of per-node data yield and per-node radio duty cycle for LWB, Dozer, and CTP over A-MAC and LPL. Figure 3.10(a) shows that all protocols but CTP + LPL deliver more than 99 % of the messages from all senders. In particular, LWB and Dozer exhibit a very high and almost identical average data yield of 99.98 %. Because of their synchronized operation, these protocols perform at their best under stable network conditions, and better than contention-based protocols like A-MAC and LPL.

Figure 3.10(b) indeed shows that LWB and Dozer achieve low average radio duty cycles of 0.43 % and 0.23 %, respectively. LWB's efficiency is due to the little control overhead to distribute schedules and allocate req slots, which accounts for only 0.05 % of a node's radio duty cycle. The average radio duty cycle we observe for Dozer also confirms that our port to the TelosB platform performs similar to the original implementation [BvRW07]. Moreover, we again observe that tree-based protocols like Dozer and CTP bias the routing load towards the receiver. For example, radio duty cycles with Dozer range between 0.04 % and 1.91 %, whereas LWB distributes the energy load more evenly, achieving per-node radio duty cycles of 0.41–0.48 %.

3.6.2   Heavy Traffic

We next consider hundreds of nodes generating relatively heavy network traffic. Applications such as data center monitoring exhibit similar aggregate traffic loads [LLL+09].

Scenario. We use the 260 nodes available on KANSEI. All nodes but the receiver act as senders and generate messages with the same fixed IPI for 4 hours. We test four different IPIs: 30 s, 20 s, 10 s, and 5 s, and compare LWB with CTP over CSMA and LPL, using wake-up intervals of 100 ms,

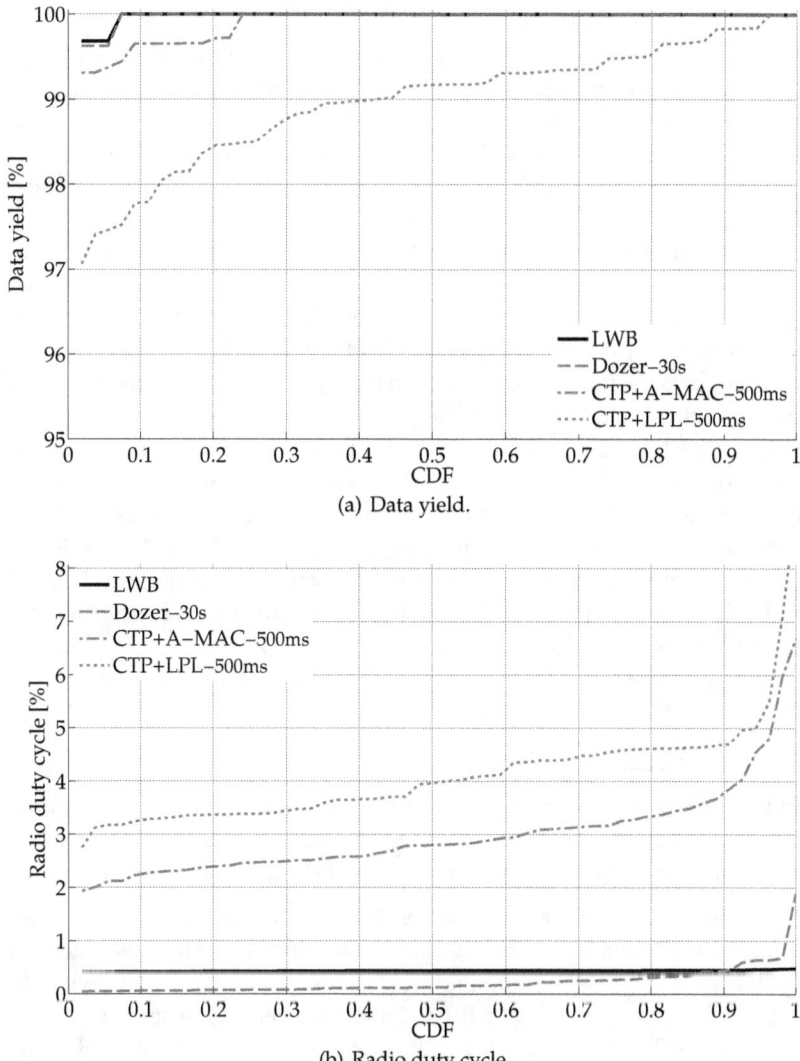

(a) Data yield.

(b) Radio duty cycle.

Figure 3.10: Per-node performance at light traffic. Synchronized protocols outperform contention-based protocols at stable network conditions.

50 ms, and 20 ms for the latter. [1] We exclude Dozer, as it is not designed for such heavy traffic: constraints on the maximum message queue size and an increased risk of collisions [BvRW07] cause significant message loss at higher traffic loads (see also Section 3.6.3).

Results. Figure 3.11 plots data yield and radio duty cycle for different IPIs. We see from Figure 3.11(a) that LWB and CTP + CSMA achieve a data yield close to 100 % across all IPIs. With LWB, the goodput at the receiver amounts to 6.1 kbps at IPI = 5 s: its synchronized operation provides sufficient bandwidth to cope with such high traffic demands. By contrast, CTP + LPL collapses at IPI = 5 s even with the shortest wake-up interval: the bandwidth provided by LPL is insufficient, leading to congestion and more than 80 % message loss. At this IPI, CTP + CSMA provides a slightly higher data yield than LWB: 99.86 % against 98.44 %. However, with LWB the average radio duty cycle at this setting is 48.25 %, thus nodes consume less than half energy for communication than with the non duty-cycled CTP + CSMA.

Figure 3.11(b) exposes the trade-off between energy costs and network capacity in CTP + LPL. A longer LPL wake-up interval may save energy, but reduces the available bandwidth. LWB constantly provides a higher network capacity and is more energy-efficient than CTP + LPL. This holds particularly for nodes in the vicinity of the receiver, which have the highest radio duty cycles with CTP + LPL as they carry the highest loads. Overall, LWB requires only 2.50 ms of radio-on time to deliver a single application message, whereas CTP + LPL needs *four* times as much.

3.6.3   Fluctuating Traffic

In this experiment, we evaluate the performance of LWB when the traffic demands change over time, which is characteristic of applications that adjust the data rates in response to external stimuli [ADB+04].

Scenario. We use 54 sender nodes on FLOCKDSN that generate messages with IPI = 60 s for 1.5 hours. During two periods of 15 minutes each, 14 spatially close nodes, grouped within "cluster A" in Figure 3.8, switch to IPI = 5 s (traffic peak 1) and IPI = 2 s (traffic peak 2), respectively. We compare LWB with Dozer and CTP over LPL. [1] In Dozer, we halve the beacon interval to 15 s and triple the queue size to 60 messages to help its performance during traffic peaks. We test 100 ms and 200 ms as LPL wake-up intervals.

---

[1]We omit inconsistent results with CTP + A-MAC. Because the current A-MAC implementation does not support multiple channels when using broadcasts, many probes collide with application packets at heavy and fluctuating traffic, which affects A-MAC's performance significantly in these scenarios.

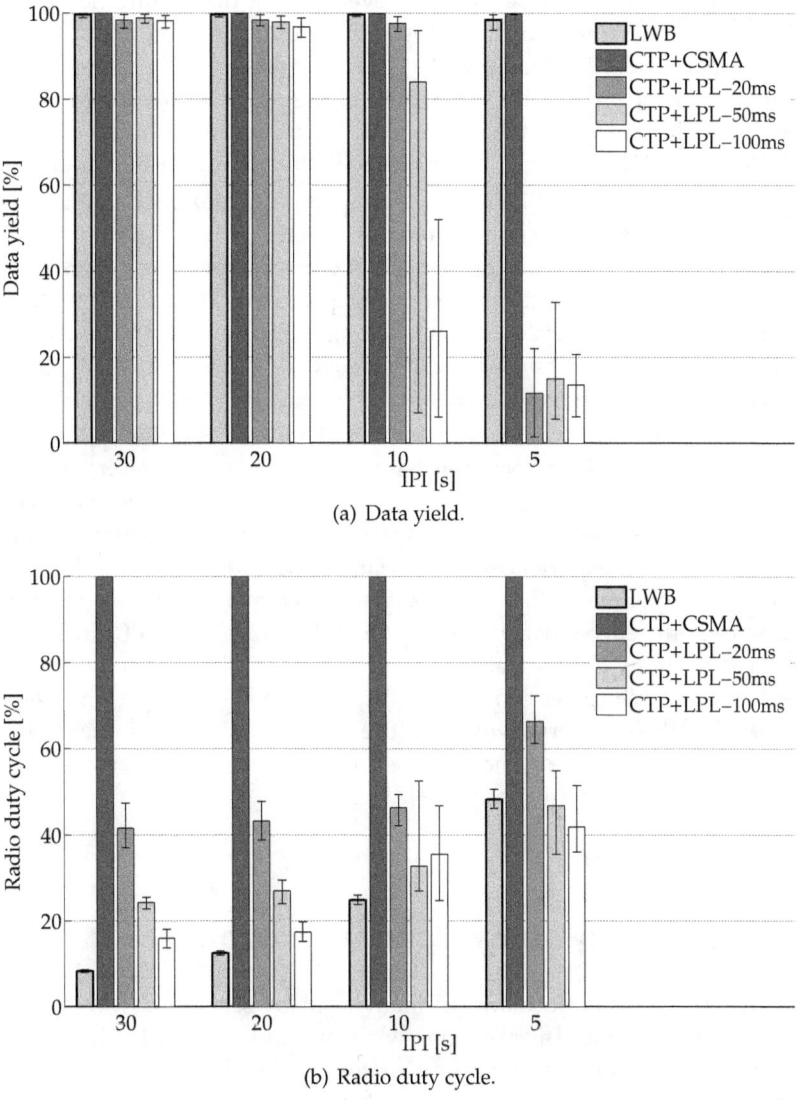

(a) Data yield.

(b) Radio duty cycle.

Figure 3.11: Performance at heavy traffic from 259 senders. Bars denote averages; error bars indicate 5th and 95th percentiles. LWB consistently provides a higher network capacity and is more energy-efficient than CTP + LPL.

Results. Figure 3.12 plots data yield and radio duty cycle over time, averaged over all sender nodes. Figure 3.12(a) shows that data yield with LWB is always close to 100 %, even during the two traffic peaks: LWB promptly reacts to the changed traffic demands and adapts the round period T. For example, the scheduler sets $T = 30\,$s when nodes generate messages with IPI = 60 s, but reduces it to $T = 7\,$s during peak 2. Figure 3.12(b) shows that the average radio duty cycle rises from 0.8 % to 7.8 % during traffic peak 2, but returns to 0.8 % once the peak is over.

Dozer and CTP + LPL lack such adaptability. Dozer's data yield is almost 100 % before peak 1, but drops severely when the traffic load increases, below 30 % during peak 2. Because Dozer sets no limit on retransmissions, messages are lost due to queue overflows at the senders. Numerous queue overflows occur during peak 2 also with the largest queue size we could fit in RAM (220 messages). With CTP + LPL the drop in data yield is less severe, and depends on the LPL wake-up interval.

More generally, Dozer's beacon interval and LPL's wake-up interval are fixed and set before operation, based on the desired trade-off between energy efficiency and network capacity. For instance, Dozer achieves the lowest average radio duty cycle of 0.35 % in this scenario. This value, however, experiences low variations when the amount of traffic changes, highlighting that Dozer's fixed functional parameters can not provide sufficient bandwidth during the traffic peaks. For LPL, the better data yield with 100 ms wake-up interval comes at the price of higher energy costs, which need to be paid also during low-traffic periods when a longer wake-up interval would suffice. Finding suitable parameters for these protocols is indeed challenging, and additional complexity is often required to adapt them at runtime [ZFM+12].

3.7 Many-to-Many Communication

We assess LWB's performance in many-to-many scenarios. These arise, for example, in control applications, where multiple senders feed different control loops running at multiple actuators [CCD+11]. Most importantly, the basic mechanisms that we use to achieve dependable communication in cyber-physical systems and exploit in Chapter 4, extensively require efficient many-to-many interactions among nodes. In the following, we observe that:

Finding 3. *LWB efficiently supports many-to-many communication without any changes to the protocol logic.*

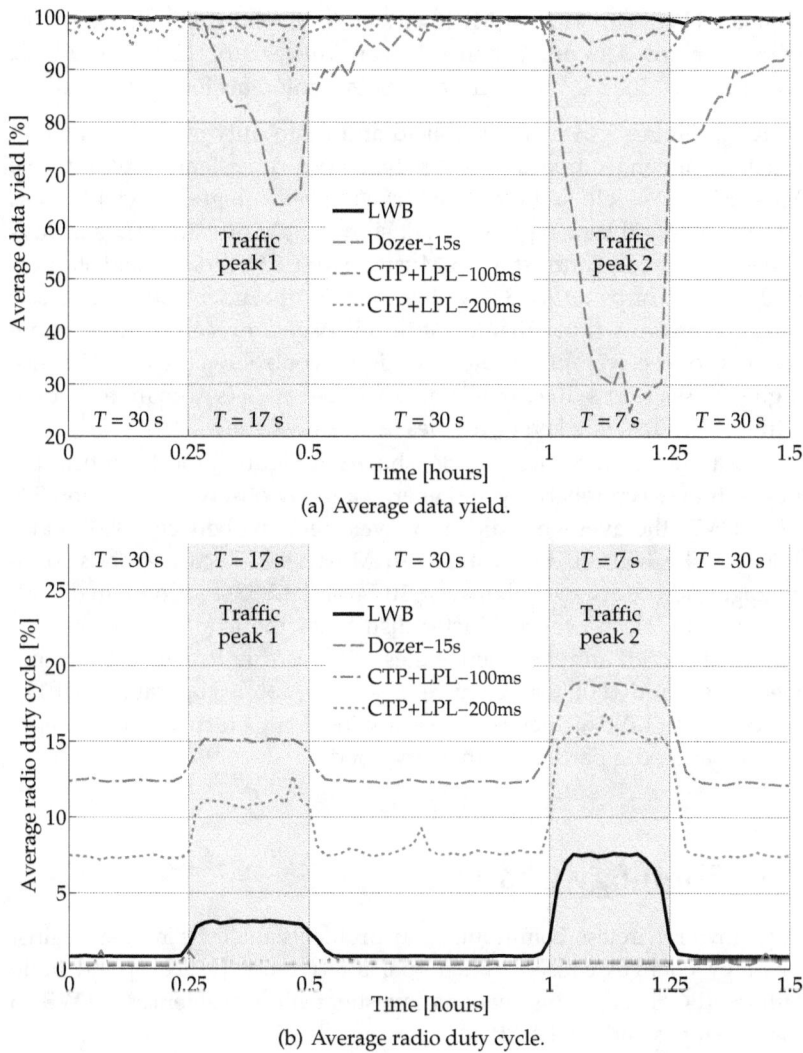

(a) Average data yield.

(b) Average radio duty cycle.

Figure 3.12: Performance as the traffic demands change. LWB balances energy costs and network capacity by adapting round period and data slot allocation.

Scenario. Out of the 90 nodes available on TWIST, we randomly pick 8 as receivers and a fraction of 0.2, 0.5, or 0.8 of the total as senders. These generate messages with IPI = 1 minute for 4 hours. We use the same LWB implementation and parameter settings as in Section 3.6. We compare LWB with Muster [MP11], a state-of-the-art routing protocol for many-to-many communication. We run Muster atop CSMA and LPL, using 500 ms, 200 ms, and 50 ms as wake-up intervals for the latter.

Results. Figure 3.13 plots data yield and radio duty cycle for different fractions of sender nodes. We see that LWB consistently outperforms Muster in data yield and radio duty cycle. The average data yield across all receivers and senders with LWB, shown in Figure 3.13(a), is always above 99.94 %. In contrast, with Muster over CSMA, data yield starts at 99.01 % and drops to 97.98 % as the fraction of sender nodes increases. Muster performs route maintenance on a sender-receiver basis; more senders (or receivers) translate into higher control overhead and hence higher message loss due to collisions. This behavior is even more evident with LPL, as this link layer provides less bandwidth.

The trends in radio duty cycle, shown in Figure 3.13(b), confirm the trade-off between reliability and energy already observed in Section 3.6. With LWB, the average radio duty cycle remains between 0.31 % and 1.06 %. The highest data yield with Muster + LPL corresponds to an average radio duty cycle between 10.14 and 12.57 %. Compared with Dozer and CTP, however, Muster distributes the load more evenly, as indicated by 5th and 95th percentiles. This is due to a load-balancing mechanism added on top of Muster's normal protocol operation [MP11]. By contrast, LWB achieves network-wide load balancing by design, because all nodes participate in every flood.

3.8   Topology Changes

Low-power wireless communication protocols must be robust against topology changes caused by external interference [LPLT10] and node failures [BGH+09]. This sections investigates the resilience of LWB to these changes and reveals that:

Finding 4. *Thanks to the absence of topology-dependent state, LWB operates efficiently also in the presence of topology changes due to external interference and node failures.*

Dozer and CTP, like most existing protocols, rely on periodic broadcasts to keep routing and synchronization state up-to-date, which is extremely costly atop contention-based link layers [PGZM12]. The high

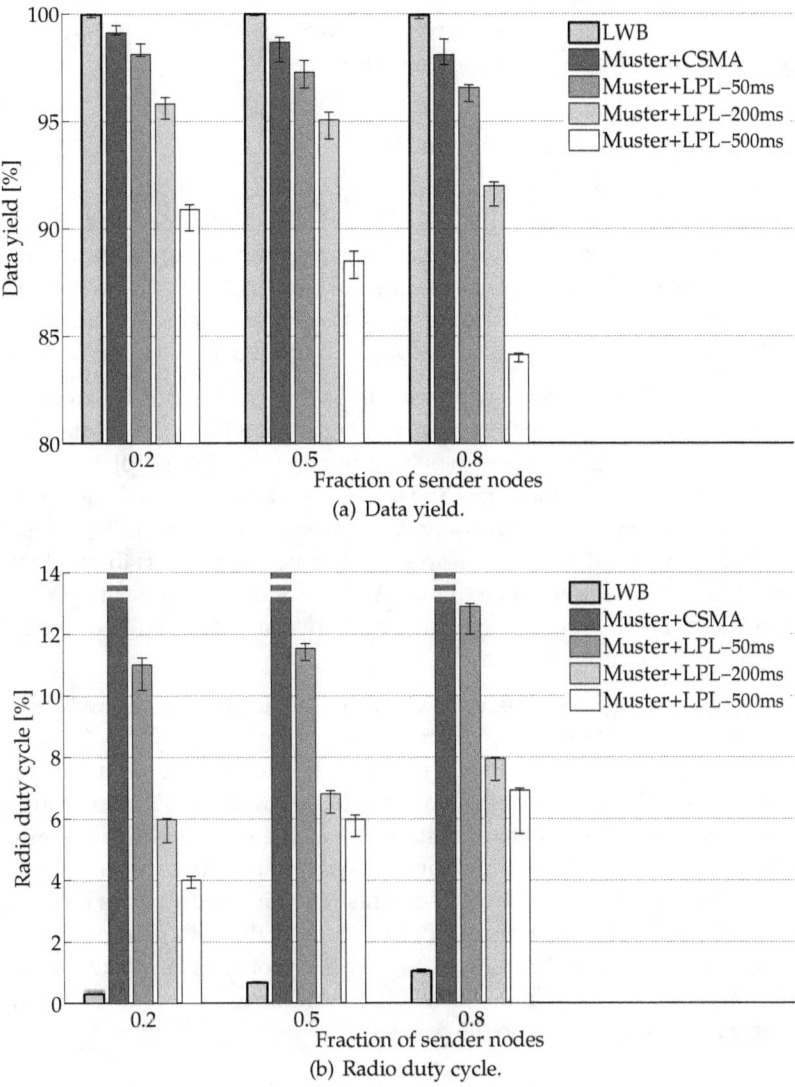

(a) Data yield.

(b) Radio duty cycle.

Figure 3.13: Performance with 8 receivers and varying fractions of sender nodes. Bars denote averages; error bars indicate 5th and 95th percentiles. LWB performs as in single-receiver scenarios, since all nodes receive all data.

efficiency of Glossy keeps to a minimum the energy cost of LWB's little control traffic, required, for example, to distribute sched messages and time-synchronize the nodes.

Section 3.9 investigates how LWB deals with extreme forms of topology change caused by node mobility.

3.8.1   External Interference

Wi-Fi interference can significantly degrade the performance of low-power wireless protocols [DDHC+10, LPLT10, SDTL10]. The use of RSSI in many protocols to detect incoming traffic and select links can lead to significantly higher radio duty cycles and lower reliability when the 802.15.4 channel used overlaps with the channels occupied by 802.11 (Wi-Fi). We assess next the robustness of LWB to Wi-Fi interference.

Scenario. We run 3-hour experiments on FLOCKDSN during working hours, letting 54 senders generate messages with IPI = 1 minute. We first use channel 26, which is most immune to Wi-Fi [SDTL10], and then channel 20, which we measure to be most affected by Wi-Fi on FLOCKDSN. We compare LWB with Dozer and CTP over A-MAC and LPL. We set the beacon interval in Dozer to 15 s to improve reactiveness to topology changes. The wake-up intervals of A-MAC and LPL are set to 500 ms and 200 ms, providing a good trade-off between data yield and energy consumption at this traffic load.

Results. Figure 3.14 plots data yield and radio duty cycle, with and without Wi-Fi interference. Bars show averages; error bars indicate 5th and 95th percentiles. Figure 3.14(a) shows that all protocols but CTP + LPL maintain high data yield also with Wi-Fi interference, averaging above 99 %. Although its average data yield slightly decreases from 99.98 % to 99.03 %, LWB shows no noticeable impact on radio duty cycle, shown in Figure 3.14(b). In contrast, the radio duty cycles increase considerably with Dozer and CTP; for example, the 95th percentile with Dozer rises from 0.60 % to 1.61 %. These protocols must adapt the routing tree to varying channel conditions, leading to higher radio activity, whereas LWB is immune to the problem.

3.8.2   Node Failures

Real deployments must deal with temporary node disconnections and persistent outages [BGH+09]. In the following, we evaluate how effectively LWB adapts to these situations.

Scenario. We run experiments on FLOCKDSN for 1.5 hours, letting 54 sender nodes generate messages with IPI = 1 minute. We adopt a similar

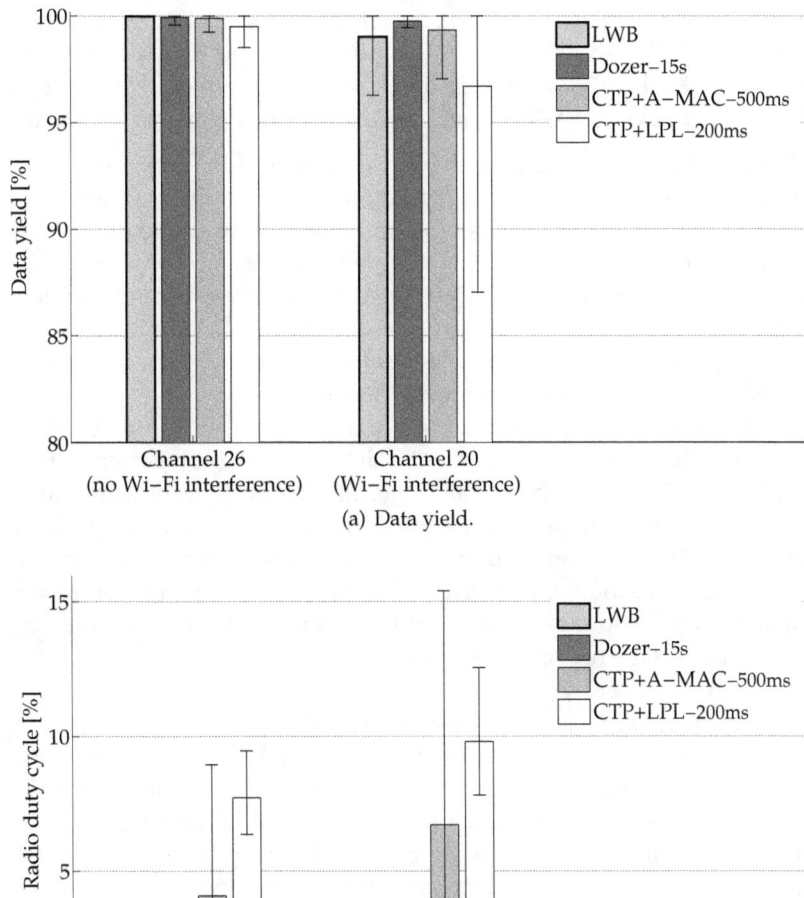

(a) Data yield.

(b) Radio duty cycle.

Figure 3.14: Performance with and without Wi-Fi interference. Bars denote averages; error bars indicate 5th and 95th percentiles. LWB delivers on average more than 99 % of messages also in the presence of Wi-Fi interference, while not affecting radio duty cycle.

scenario as Gnawali et al. [GFJ+09]: after 15 minutes, we turn off 8 nodes in the vicinity of the receiver (cluster B in Figure 3.8). We turn them on again after 15 minutes, and repeat the off-on pattern after 30 minutes. We compare LWB with Dozer and CTP over CSMA and A-MAC. Because the traffic load is the same as in the previous experiment, we set again Dozer's beacon interval to 15 s and A-MAC's wake-up interval to 500 ms.

Results. Figure 3.15 plots data yield and radio duty cycle over time, averaged over all functional nodes. In the first 15 minutes, all protocols deliver more than 99 % of messages and have a stable radio duty cycle. When 8 nodes are turned off, we observe no noticeable change in LWB's data yield, shown in Figure 3.15(a), since its route-free operation renders state reconfigurations unnecessary. The only effect is that the host realizes that it receives no more data messages from streams generated by the 8 failed nodes and eventually removes these streams.

The removal of inactive streams reduces the number of allocated data slots, which makes the average radio duty cycle slightly decrease from 0.83 % to 0.72 %. When the failed nodes recover, they turn the radio on to synchronize again, causing the short increases in radio duty cycle in Figure 3.15(b). As soon as the scheduler receives the first stream request, it reduces the round period to $T_{min} = 1$ s to make nodes join faster. In less than 20 s all 8 nodes are again part of the bus, and after 1 minute the scheduler sets the round period back to 30 s.

Although it achieves the lowest average radio duty cycle of 0.35 % when the topology is stable, Dozer's average data yield drops below 96 % when nodes are removed. A slight dip is also visible with CTP, even when using CSMA as the link layer. After a failure, both protocols update the routing tree, which generates more control traffic and thus higher radio activity, as visible from the increase in radio duty cycle. This process is also prone to temporary inconsistencies such as routing loops. These factors all concur to message losses. With no routing state to update, LWB keeps delivering messages reliably without an increases in energy costs, as long as the network is connected. Moreover, the scheduler effectively detects and removes streams of failed nodes, saving bandwidth and energy if failures persist, and adapts the round period to accelerate rejoining of nodes after temporary outages.

3.9   Mobility

We evaluate LWB's performance in the presence of mobile nodes, a scenario that proved to be extremely challenging in multi-hop low-power wireless networks [DC09, KLW+09, LKA+10]. At the same time, emerging

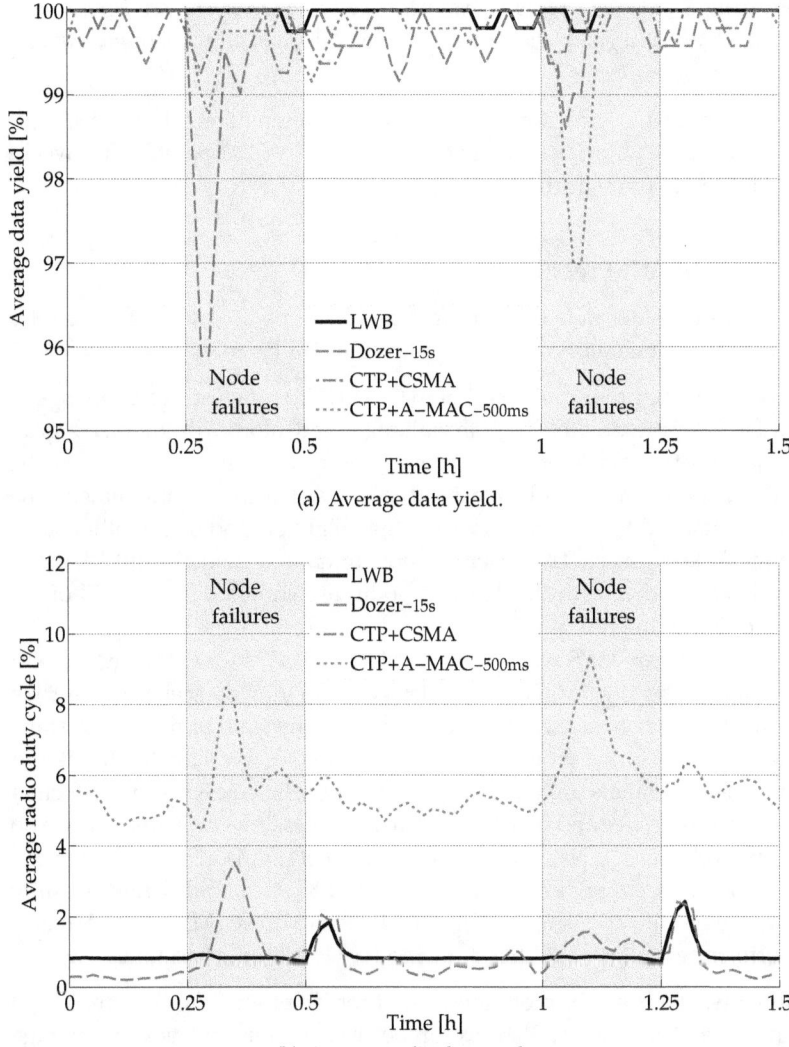

(a) Average data yield.

(b) Average radio duty cycle.

Figure 3.15: Average performance while 8 nodes concurrently fail. When nodes recover, LWB reduces the round period to make them quickly join the bus.

real-world applications increasingly rely on the ability to attach sensor nodes to mobile entities [BEP+06, CLBR10, DEM+10]. Our experimental results indicate that:

Finding 5. *LWB supports mobile nodes acting as receivers, senders, or both without any changes to the protocol logic and performance loss compared with the static network case.*

We study next specific settings modeling different mobile applications, and conclude with a real-world assessment of LWB's operation in a week-long test at our institution.

3.9.1   Mobile receiver

We first consider data delivery to a mobile receiver, a setting frequently found in participatory sensing applications [BEP+06].

Scenario. We program one robot in CONETIT to move at a constant speed of 1 m/s (approximately human walking speed) along a predefined zigzag trajectory that starts at one corner of the testbed area and ends at the opposite corner. Every run lasts 30 minutes. During the movement, the node attached to the robot is within the neighborhood of any other static node at least once. The other 4 robots remain at their default locations and act together with the 21 static nodes as senders, using an IPI of 4 s, 2 s, or 1 s in different runs.

We compare LWB with BCP + CSMA and CTP + CSMA, using default parameter settings. CTP is not designed for mobile scenarios, but we consider it nevertheless to understand how the state of the art for static networks performs when the receiver is mobile. We start the robot and our measurements after 15 minutes, giving CTP enough time to form a stable routing tree and BCP to build up backpressure gradients. We use no duty-cycled link layer: most existing protocols supporting mobile nodes do not target energy efficiency [MSKG10, LKA+10], and would possibly require modifications to existing link layers to do so [MSKG10]. We also perform several runs with a static receiver to obtain a baseline.

Results. Figure 3.16 plots the results for different IPIs. Figure 3.16(a) shows that the same LWB prototype used so far achieves an average data yield above 99.94 % also when the receiver moves. This is because LWB keeps no topology-dependent state: as long as the network is connected, LWB is oblivious to topology changes. In contrast, BCP and CTP deliver fewer messages when the receiver moves: they need to constantly reconfigure state that depends on a node's current neighbors and connectivity. Although the degree of mobility is fairly limited, this already suffices to affect the performance of both protocols.

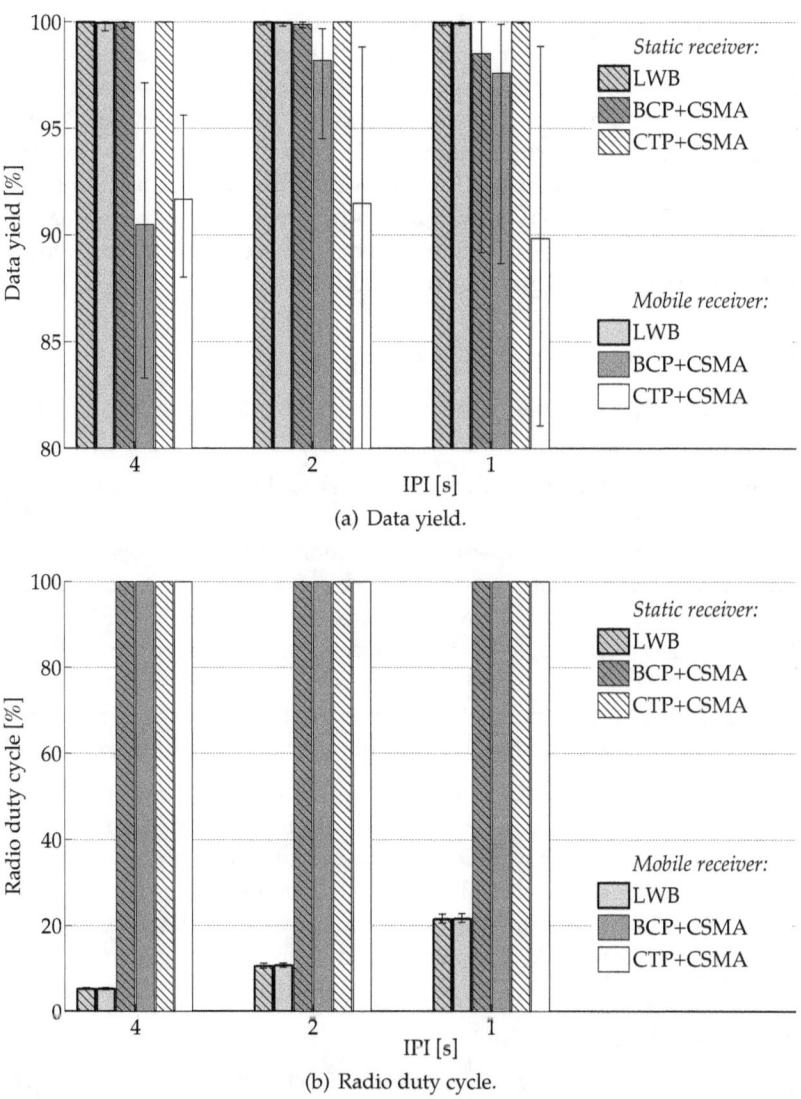

(a) Data yield.

(b) Radio duty cycle.

Figure 3.16: Performance when the receiver is either static or mobile. Bars denote averages; error bars indicate 5th and 95th percentiles. Unlike other protocols, LWB has the same data yield and radio duty cycle regardless of whether the receiver moves or not.

A deeper look reveals that with receiver mobility CTP delivers consistently around 90 % of the messages almost regardless of data rates. Instead, BCP's data yield peaks at 98 % with IPI = 2 s. We conjecture that the optimal parameter setting in BCP is sensitive to traffic load. In contrast, LWB is equally efficient in both static and mobile networks with the same configuration, also in terms of radio duty cycle: Figure 3.16(b) shows that LWB's performance in a mobile setting is similar to that in static networks regardless of the traffic load. By contrast, running BCP and CTP atop a duty-cycled link layer likely results in poorer delivery performance, as the reduced network capacity would make these protocols adapt more slowly to the ever-changing network conditions.

3.9.2   Mobile Senders and Mobile Receiver

We look at a typical setting where mobile senders deliver data to a mobile receiver via a stationary infrastructure [KLS⁺10].

Scenario. Four robots on CONETIT act as mobile senders, generating messages with IPI = 1 s; one robot acts as the mobile receiver. The robot trajectories and experiment duration are as in Section 3.9.1. The remaining 21 static nodes generate no messages and form a stationary relay backbone. We compare LWB with BCP + CSMA and CTP + CSMA. Although similar solutions are employed in settings akin to ours [DMT⁺11], these protocols are not expressly designed for mobile senders. Unfortunately, we could not gain access to a reliable implementation of an alternative baseline conceived for such scenarios.

Results. Figure 3.17 depicts the results. Overall, the performance is consistent with the mobile receiver case discussed above. To leverage mobile senders through a static infrastructure, LWB requires no changes to the protocol logic. Specifically, LWB achieves an average performance of 99.98 % in data yield and 0.84 % in radio duty cycle. The latter figure is lower than in Section 3.9.1, because now only four nodes generate data.

3.9.3   Real-World Trial

Finally, we run a week-long experiment at ETH to study LWB in a longer-term setting involving many-to-many and one-to-many traffic, changes in traffic demands and active nodes, and mobile nodes acting as senders and receivers. Such setting would currently require two network-layer protocols (e.g., Muster and Trickle [LPCS04]) atop a link-layer protocol. LWB provides all required features in a single protocol logic.

Scenario. We use 5 battery-powered nodes carried by people around FLOCKDSN on 7 consecutive days during working hours. Nodes in

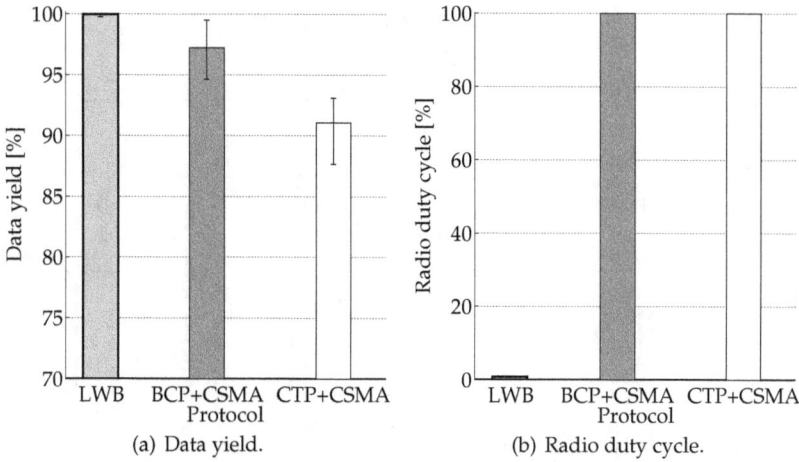

(a) Data yield.

(b) Radio duty cycle.

Figure 3.17: Performance with a mobile receiver and 4 mobile sender nodes. Bars denote averages; error bars indicate 5th and 95th percentiles.

FLOCKDSN form a static infrastructure. People carrying the nodes entail less structured movement than in the previous experiments, mimicking mobility of real-world applications [CLBR10]. All mobile nodes act as both senders and receivers, generating messages with IPI = 5 minutes. All (static) nodes in FLOCKDSN also generate messages at the same rate.

To induce changes in the traffic demands and set of active nodes, people switch their node off when they leave (e.g., after work). They switch their node on again when they come back, eventually reconnecting the node to the bus. One mobile node, named B, plays a special role: the person can press its user button to trigger a second high-rate stream at IPI = 1 s. When the other 4 mobile nodes M_1–M_4 recognize this, they generate such high-rate stream as well. When the person presses the button again, B cancels the high-rate stream and so do M_1–M_4.

We note that pressing the button triggers the application on B to issue a stream add request with IPI = 1 s starting at the current time. To inform the host, B piggybacks on application data messages or transmits the stream request in a req slot. After receiving a stream acknowledgment from the host, B eventually starts sending more messages in the data slots allocated to the second high-rate stream. These messages have an application field set that signals M_1–M_4 to issue a similar stream add request. The host then allocates further data slots to these high-rate streams and adapts the round period, for example, reducing it from 30 s to 11 s when all 5 mobile nodes are active. Pressing the button on B again triggers a remove request to cancel the high-rate stream. This initiates a reverse chain of actions

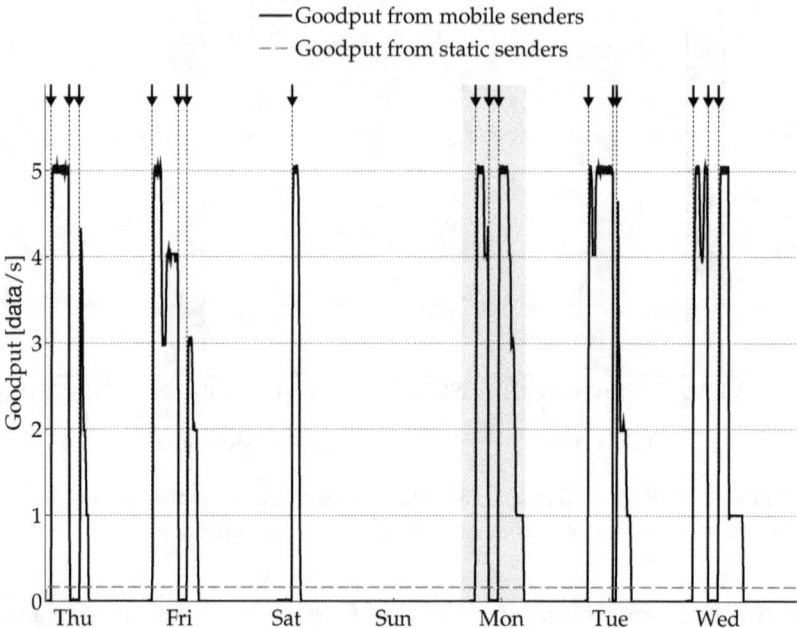

Figure 3.18: Goodput at a static node in a real-world trial, in data messages received per second. Arrows indicate when the button is pressed on B.

at the host and the other mobile nodes—eventually all active nodes emit again only one low-rate stream.

Results. Figure 3.18 shows the goodput at a static node throughout the entire week. The arrows on top indicate when a button is pressed on node B. We see from the dashed line that the node receives a constant amount of data messages from the static senders, because these nodes always generate data at low rate. The solid line shows instead that the number of data messages received from the mobile nodes varies significantly over time. This is because the mobile nodes start and stop generating an additional stream with IPI = 1 s when a button is pressed on B. Moreover, the number of received data messages depends on the number of currently active mobile nodes, and decreases, for example, when people turn off their nodes at the end of a day.

Figure 3.19 shows goodput and radio duty cycle of the 5 mobile nodes during a 14 hours excerpt of our measurements corresponding to the grey area in Figure 3.18. At about 10 AM, node B triggers the high-rate stream. All mobile nodes are running at this time besides M_2, which is off the very moment the traffic peak begins. It also starts generating high-rate messages as soon as it becomes active. This corresponds to the reception

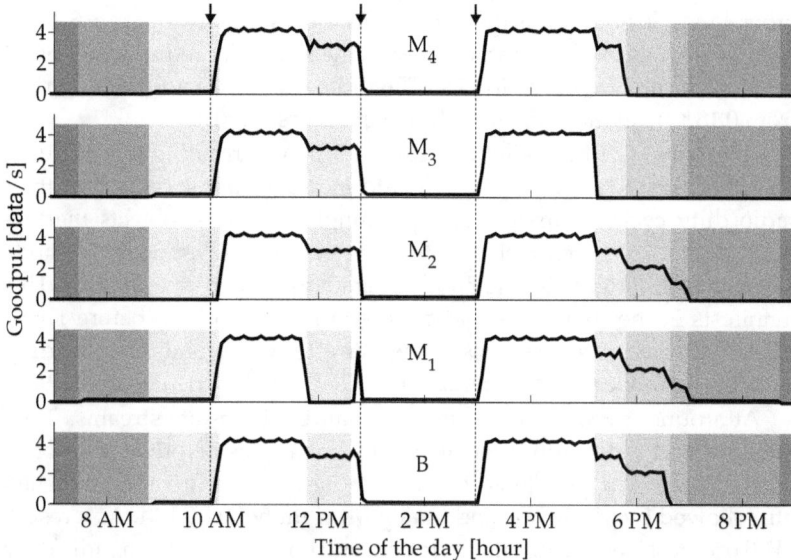

(a) Goodput at the mobile nodes, in data messages received per second.

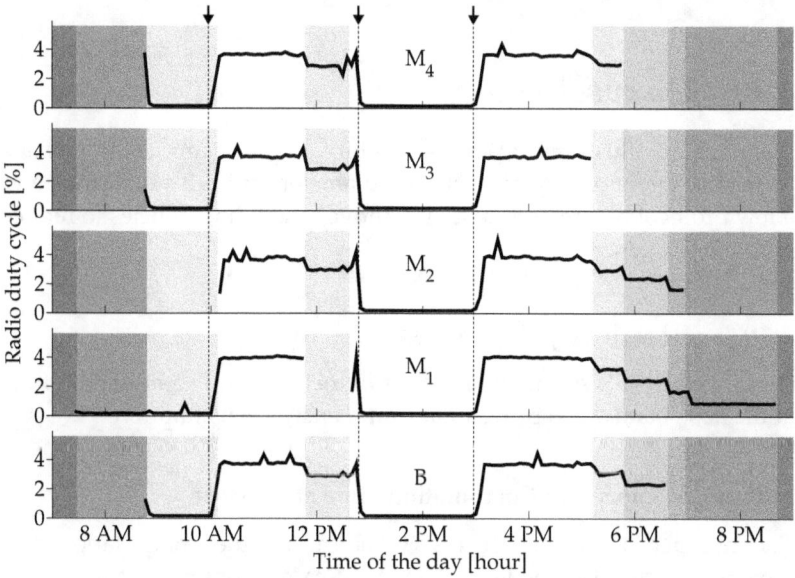

(b) Radio duty cycle of the mobile nodes.

Figure 3.19: Goodput and radio duty cycle of 5 mobile nodes in a real-world trial. Arrows indicate when the button is pressed on B; areas are lighter when more mobile nodes are active.

of slightly more than 4 data messages per second, one from each of the other 4 mobile nodes plus the low-rate traffic, as shown in Figure 3.19(a). As a consequence of the traffic increase, the scheduler reduces the round period T from $T_{max} = 30$ s to 11 s. The radio duty cycle accordingly rises from 0.18 % to about 3.78 %, as shown in Figure 3.19(b).

At around 12 PM, node M_1 is turned off. As a result, the goodput at the other mobile nodes is about 3 data messages per second, and their radio duty cycle decreases correspondingly: the host detects that M_1 disconnected and then reclaims its active streams. Close to 1 PM, node M_1 restarts, but right after that node B removes all high-rate streams. This manifests in the short-lived peak shown in Figure 3.19(a) before 1 PM. Nodes only generate messages at low rate afterwards, and the radio duty cycle of all mobile nodes drops again to 0.18 %, as shown in Figure 3.19(b).

At around 3 PM, node B triggers again the high-rate streams. Both goodput and radio duty cycle increase similarly as before at all mobile nodes. After 5 PM, people start leaving: node M_3 is the first to be switched off, followed by M_4, B, M_2, and finally M_1 right before 9 PM. As a result, LWB progressively adapts its operation, and both goodput and radio duty cycle decrease in a step-wise fashion.

3.10   Discussion

This section illustrates LWB's scalability as the number of streams increases, the impact of the network diameter on LWB's efficiency and a few protocol parameters, and alternative scheduling policies to reduce end-to-end latency.

3.10.1   Scalability Properties

We discuss LWB's scalability in terms of memory and computation time at the host, bandwidth provisioning, and energy consumption.

3.10.1.1   Memory and Computation Time at the Host

The number of active streams S determines the computation and memory overhead at the host. The worst-case computation time in our experiments is 49 ms with 259 streams (see Section 3.6.2). Memory scales linearly with the number of active streams S; our LWB prototype uses 13 bytes per stream. Nevertheless, memory and computation costs are paid only at the current host, and proved to be affordable with several hundreds of streams on TelosB nodes.

3.10.1.2   Bandwidth Provisioning

Bandwidth provisioning also scales linearly with the number of active streams S, as LWB's bus-like operation prevents spatial bandwidth reuse. Depending on the number of streams and their IPIs, different solutions may perform better. In a sense, we hit this point in the KANSEI experiments in Section 3.6.2, where CTP + CSMA slightly outperforms LWB in data yield at IPI = 5 s. This, however, comes at 100 % radio duty cycle—a possible, yet rarely affordable design choice in real-world applications.

Our evaluation demonstrates that LWB always outperforms all asynchronous duty-cycled protocols we consider, and closely matches the performance of Dozer in a scenario particularly suited to the latter. We achieve this result with a single design that encompasses multiple traffic patterns and seamlessly supports node mobility in connected networks. We maintain these features are worth the non-optimal bandwidth scaling LWB may show in specific settings. Nevertheless, hierarchies of buses may improve LWB's scalability on networks with several hundreds or thousands of nodes. By assigning different wireless channels to disjoint sets of nodes, nodes can grouped into clusters (i.e., low-level buses), while several nodes from each cluster connect also to a shared high-level bus to exchange data among clusters.

3.10.1.3   Energy Consumption

Energy consumption also scales linearly with the number of active streams S. We now show this by introducing a model that provides a *pessimistic* estimation of a node's radio duty cycle in *connected* networks.

Estimation of radio duty cycle.   Our model is pessimistic because it assumes that nodes keep the radio on for the entire duration of a communication slot. Glossy instead turns the radio off as soon as the node has transmitted N_{tx} times during a flood (see Chapter 2), which typically happens before the end of a slot. By connected we mean that a node does not miss sched messages for more than three consecutive times, which would make it turn its radio on until a new sched reception (see Section 3.2.5). Due to the high reliability of Glossy, this happens with an extremely high probability unless nodes fail or the network partitions.

For simplicity, our model considers sender nodes generating data messages with the same, constant rate. We thus assume that S periodic streams generate data with common interval $IPI_1 = \cdots = IPI_S = IPI$. For this scenario, we can rewrite (3.1) as $T_{opt} = D \times IPI/S$, and according to (3.2) the host computes the round period T as

$$T = \left\lfloor \min\left(T_{max}, \max\left(D \times \frac{IPI}{S}, T_{min}\right)\right)\right\rfloor \qquad (3.6)$$

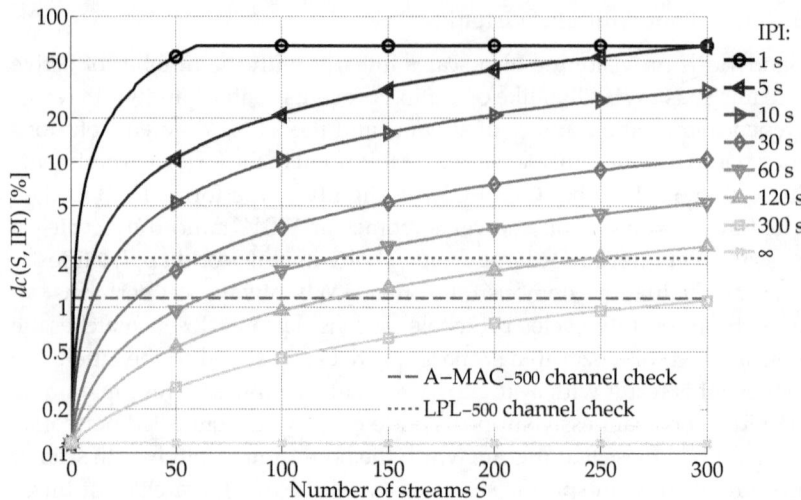

Figure 3.20: Estimated radio duty cycle dc, for several combinations of number of streams S and IPI. *Note the logarithmic scale on the y-axis.*

LWB provides three types of communication slots (see Figure 3.2): sched, data, and req slots. The overall radio duty cycle dc of a node is thus the sum of their individual contributions: $dc = dc_{sched} + dc_{data} + dc_{req}$:

- dc_{sched}. Every round period T the host allocates two sched slots of length T_{sched}: $dc_{sched} = 2 \times T_{sched}/T$.

- dc_{data}. Every round period T the host allocates on average $\min(T \times S/\text{IPI}, D)$ data slots of length T_{data}, resulting in a radio duty cycle $dc_{data} = \min(T \times S/\text{IPI}, D) \times T_{data}/T$.

- dc_{req}. Under stable traffic conditions, the host allocates one req slot of length T_{req} every T_r: $dc_{req} = T_{req}/T_r$.

The overall radio duty cycle dc is thus a function of S and IPI:

$$dc\,(S,\, \text{IPI}) = 2 \times \frac{T_{sched}}{T} + \min\left(\frac{T \times S}{\text{IPI}},\, D\right) \times \frac{T_{data}}{T} + \frac{T_{req}}{T_r} \qquad (3.7)$$

Figure 3.20 plots the estimated radio duty cycle dc with up to 300 active streams that have IPI between 1 s and ∞, the latter representing the extreme case where they generate no data messages.

Control overhead. We first use (3.7) to estimate the minimum *control overhead* required by LWB to operate. This corresponds to streams generating no data messages (i.e., IPI → ∞), and thus to zero duty cycle due to data slots (i.e., $dc_{data} \to 0$). The resulting duty cycle when using the default LWB configuration of Table 3.1 is

$$dc(S, \infty) = 0.1167\% \qquad (3.8)$$

This value corresponds to the lowest line in Figure 3.20.

To put this value into perspective, we also plot the minimum control overhead of A-MAC and LPL, required to periodically check the channel. We set the wake-up interval to 500 ms in their TinyOS implementations and measure the radio duty cycle when no (application or routing) traffic is generated. We measure radio duty cycles of 1.16 % for A-MAC and 2.19 % for LPL, values one order of magnitude higher than for LWB. Figure 3.20 shows that at the same radio duty cycles, for example, LWB supports more than 50 and 100 streams with IPI = 60 s, respectively.

Scalability. Figure 3.20 shows also that radio duty cycle increases linearly with the number of active streams S, when they generate data messages (i.e., IPI < ∞). Radio duty cycle reaches its upper bound dc_{max} when the system saturates, because the host allocates the maximum number of data slots $D = 60$ at the minimum round period $T_{min} = 1$ s:

$$dc_{max} = 2 \times \frac{T_{sched}}{T_{min}} + D \times \frac{T_{data}}{T_{min}} + \frac{T_{req}}{T_r} \approx 63\% \qquad (3.9)$$

Nevertheless, our model confirms that the same LWB configuration achieves radio duty cycles below 1 % in medium-scale networks with light traffic but is also highly energy efficient in networks with several hundreds of nodes generating heavy traffic. For instance, it estimates a radio duty cycle of 13.72 % when 259 streams generate data messages with IPI = 20 s, confirming the maximum radio duty cycle of 13.63 % measured on KANSEI during the testbed experiments in Section 3.6.2.

3.10.2   Impact of Network Diameter

The time taken for a Glossy flood to cover the entire network depends on the network diameter (see Chapter 2). Because LWB uses only Glossy floods for communication, its efficiency decreases in deep networks that span several tens of hops, and other approaches may perform better [KPC+07]. In such scenarios, LWB's performance could be improved by integrating channel diversity and parallel pipelining into Glossy, as done in Splash [DCL13].

In particular, the length of sched slots T_{sched}, data slots T_{data}, and req slots T_{req} must be sufficient for a Glossy flood to cover the entire network. Our setting in the evaluation is sufficient for networks whose physical topology spans at most 7 hops. However, networks may be

Protocol	Data yield	Radio duty cycle
LWB	99.97 %	0.81 %
LWB-long-slots	99.98 %	0.83 %

Table 3.4: Average LWB performance with two different settings for the length of communication slots.

longer and it may be difficult to determine in advance the network diameter. In these situations, LWB users need to conservatively increase T_{sched}, T_{data}, and T_{req}.

We study how this affects LWB's performance through 3-hour experiments on FLOCKDSN with 54 senders that generate messages with IPI = 1 minute.   Besides the default parameter setting, we test a configuration called *LWB-long-slots* that doubles the values for T_{sched}, T_{data}, and T_{req} to support network diameters of up to 14 hops. All other parameters retain their original values. As shown in Table 3.4, we find that the average radio duty cycle increases only by 0.02 %: using Glossy, nodes turn their radios off after transmitting N_{tx} times during a flood (see Chapter 2), which typically happens before the end of a slot already with the default parameter setting. Data yield is marginally affected: LWB-long-slots delivers 99.98 % of data against 99.97 %.

Longer slots, however, translate into fewer available slots per round, and thus into a decrease in bandwidth. For example, based on (3.1), we conclude that the default parameter setting would support S = 300 streams with IPI = 5 s or higher without saturating the network, whereas LWB-long-slots would sustain at most IPI = 10 s from 300 streams. In the applications we target, however, this bandwidth still largely suffices.

3.10.3   Alternative Scheduling Policies

The scheduling policy in Section 3.3 aims at energy savings while still being responsive to changes in the network. To do so, it trades message latency for energy efficiency. Although this choice fits the applications we target, others may afford to spend some energy for smaller latencies.

To cater for different performance requirements, it is sufficient to change the scheduling policy.   We provide two simple alternative scheduling policies that aim primarily at minimizing message latency. The first policy, *LWB-low-latency*, adapts the round period T such that the next round occurs immediately after the generation of new messages. The second policy, *LWB-fixed-period*, fixes $T = T_{\text{min}}$. Both policies use the same stream handler and slot allocation functionality as the original policy (see Section 3.3). We assess their performance based on 3 hours experiments

Protocol	Data yield	Radio duty cycle	End-to-end latency
LWB	99.98 %	1.40 %	11.13 s
LWB-low-latency	99.83 %	1.44 %	1.19 s
LWB-fixed-period	99.99 %	1.94 %	1.23 s
Dozer-30 s	98.42 %	0.19 %	31.82 s
CTP + A-MAC-500 ms	99.80 %	4.16 %	1.73 s
CTP + LPL-200 ms	98.97 %	6.99 %	0.42 s

Table 3.5: Average performance of three LWB scheduling policies versus Dozer and CTP over A-MAC and LPL.

on TWIST, where all nodes but a receiver generate data at IPI = 1 minute; nodes use transmit power -15 dBm. In addition to the usual performance metrics, we measure end-to-end latency by timestamping messages at the sender. We compare LWB with Dozer and CTP over A-MAC and LPL.

Table 3.5 shows the results. We see that the two alternative policies achieve average message latencies in the order of 1 s, similar to CTP + A-MAC and CTP + LPL. This comes at a marginal increase in energy costs for LWB-low-latency, whereas LWB-fixed-period shows a higher radio duty cycle due to the overhead for distributing sched messages every T_{min} = 1 s.

Worth noticing is that in LWB the logic to trade performance requirements resides at a single place, the scheduler, whereas most other protocols may require non-trivial modifications in various places. Users can thus easily implement custom LWB schedulers using well-defined interfaces. The investigation of scheduling policies for LWB constitutes an interesting area for further research, possibly by applying established concepts from the real-time and embedding computing literature.

3.11   Related Work

Flooding has long been considered too expensive for regular communication in low-power wireless networks. Nevertheless, a few protocols exploit the robustness of flooding for routing while reducing collisions and energy costs, mostly by using random delays or by completely suppressing retransmissions [Mar04, YZLZ05, ZF06]. Different from LWB, these protocols keep substantial topology-dependent state, which increases their control overhead and sensitivity to link changes.

Completely contrary to LWB's flooding-based approach, the Broadcast-Free Collection Protocol (BFC) [PGZM12] avoids costly broadcasts in the presence of duty-cycled link layers altogether.

Targeting low-rate data collection at a single receiver, BFC achieves significant energy savings in comparison with CTP even under poor connectivity conditions, which comes at the price of higher delays when forming the collection tree initially. By contrast, LWB is applicable to a wider range of scenarios and bootstraps as fast as CTP.

Low-power wireless stacks typically feature a link, network, and possibly a transport layer. The latter may, for instance, adapt the sender rates to counteract congestion [PG07]. At the network layer, routing protocols construct multi-hop paths based on some cost metric [AY05]. Efficient solutions exist that tackle the single-receiver [GFJ+09], multi-receiver [MP11], mobile receiver [LKA+10, MSKG10], and mobile senders [GLJ11] case. In addition, Trickle-based protocols provide reliable network-wide data dissemination [LPCS04]. At the link layer, the many protocols available can be divided into contention-based and TDMA-based [Lan08]. The former, sender-initiated [PHC04] or receiver-initiated [DDHC+10], better tolerate topology changes. TDMA-based protocols like Dozer [BvRW07] enable higher energy savings. WirelessHART, an open standard for industrial process monitoring and control, uses TDMA to approach deterministic communication [SHM+08]. DRAND is a distributed, randomized TDMA slot assignment algorithm operating on a node's two-hop neighborhood [RWMX06].

The virtual single-hop connectivity provided by the LWB strongly differentiates it from existing TDMA protocols. While existing approaches allocate time slots to (possibly multiple non-interfering) sender-receiver pairs, LWB requires no information about the network topology and computes a global schedule solely based on the application requirements, such as desired bandwidth and end-to-end communication delay. Moreover, LWB replaces the standard network stack with a single-layer solution. Our experimental results demonstrate that LWB supports efficient and reliable many-to-one, one-to-many, and many-to-many traffic in both static and mobile scenarios. On top of this, it exploits and integrates Glossy's accurate global time synchronization, which is key in many real-world applications [ADB+04, WALJ+06].

LWB's design is inspired by prior work on fieldbus-based communication protocols [KG93]. Intended for distributed real-time control applications, these protocols primarily focus on providing predictable transmissions and guaranteed timeliness. Different from these protocols, LWB must cope with the unreliability of low-power wireless communications and the resource limitations of the employed devices, particularly in terms of bandwidth, energy, and memory.

3.12   Summary

Many emerging low-power wireless applications feature multiple traffic patterns and mobile nodes in addition to static ones, but existing communication protocols support only individual traffic patterns in dedicated network settings. LWB overcomes this limitation by using exclusively Glossy floods for communication, thereby making all nodes in the network potential receivers of all data. As a result, LWB inherently supports one-to-many, many-to-one, and many-to-many traffic. Since LWB also keeps no topology-dependent state, it is more resilient to external interference and node failures than prior approaches, and seamlessly caters for node mobility without any performance loss. Our experimental results confirm LWB's versatility and superior performance across a variety of scenarios.

LWB thus provides a unified solution for a broad spectrum of applications, ranging from traditional data collection to emerging control and mobile scenarios. This is also demonstrated by several initial works based on LWB. For example, LWB is used as the communication support during a two-week deployment of a reliable nurse call system for patients with Duchenne muscular dystrophy [ZFL+13]. Hewage et al. provide preliminary results showing that LWB may efficiently support the Transmission Control Protocol (TCP), one of the core protocols of the Internet protocol suite (IP), potentially enabling LWB-based Internet of Things systems [HV13]. In the next chapter, we present VIRTUS, a protocol that builds on LWB and provides virtual synchrony guarantees.

4

VIRTUS: A Wireless Bus with Virtual Synchrony Guarantees

As discussed in the Introduction of this thesis, applying established designs of dependable distributed systems to cyber-physical systems is often not possible, as these require guarantees that existing communication protocols for multi-hop low-power wireless networks do not provide. Such guarantees include, for example, well-defined message delivery orderings that facilitate the implementation of replicated functionality, as well as failure handling mechanisms operating with respect to both node crashes and message omissions [Sch90, KDK$^+$89].

In fact, existing low-power wireless protocols typically operate in a best-effort manner, their design being optimized towards non-functional properties, such as energy consumption [AY05]. LWB, for example, achieves data yields above 99 % in all scenarios evaluated in Chapter 3 but cannot provide any guarantee on message delivery. Nevertheless, the characteristics of typical low-power wireless networks make providing even simple communication guarantees extremely difficult. According to our conjecture in page 6, however, LWB's bus-like operation should enable the design of dependable cyber-physical systems.

Virtual synchrony. The virtual synchrony [Bir05] model for distributed computation may be one of the designated technologies to underpin dependable cyber-physical systems. Two key concepts concur to create virtually synchronous executions.

First, virtual synchrony entails a notion of *group*: a set of processes exchanging messages originated at one node in the group and addressed to all other group members. The group membership is reflected in a data structure called *view*—maintained at every group member—which reports information on the nodes in a group at a given point in time. As group members fail and new nodes possibly join, a virtually-synchronous system must ensure that the corresponding changes in the view occur in the *same order* at all group members.

Second, message exchanges must occur according to a notion of *atomic multicast*. This grants applications the guarantee that messages are delivered to either *all* or *no* group members. Moreover, message deliveries must happen in the *same order* at all group members—a feature called *total order*.

As a result of these two mechanisms, applications run under the illusion that the distributed executions are synchronous and fault-free, although the underlying interactions are way more complex. This greatly eases the design of dependable distributed systems, for example, based on replication techniques [Sch90], as every replica sees the same events— changes in the group membership and reception of messages—in the same order. Birman et al. summarize virtual synchrony as follows [BJ87]:

> It will appear to any observer—any process using the system—that all processes observed the *same events* in the *same order*. This applies not just to message delivery events, but also to failures, recoveries, and group membership changes.

Contribution and road-map. This chapter presents VIRTUS, a virtually-synchronous inter-process messaging layer we conceive for typical resource-constrained CPS platforms. To provide virtual synchrony, VIRTUS combines a dedicated *atomic multicast service*—delivering messages reliably and with total order—with a custom *view management service*—managing group changes as nodes fail or join. This renders applicable a vast portion of the existing literature on dependable distributed systems [Sch90], enabling formally-proven dependable operation of cyber-physical systems.

After illustrating in Section 4.1 the system model we base this work upon, in Section 4.2 we briefly highlight the features required for virtual synchrony (e.g., reliable and totally-ordered multicast) that are missing in LWB, which we use as a foundation for VIRTUS. Section 4.3 describes the functionality VIRTUS adds to provide atomic multicast and view management, along with formal proofs that our design does provide virtually-synchronous executions. We then describe in Section 4.4 how we complement virtual synchrony in VIRTUS with further delivery policies, and report implementation details for our target platform in Section 4.5.

Virtual synchrony comes at a cost. Based on extensive real-world experiments on two wireless sensor testbeds, we show in Section 4.6 that our VIRTUS implementation provides virtual synchrony at a marginal cost compared with LWB's best-effort operation. For example, message latency and energy consumption increase only by 1 % and 11 %, respectively. We also report on the impact of different settings of the VIRTUS parameters, demonstrating the ease to fine-tune the system.

To the best of our knowledge, we are the first to offer formally-proven virtual synchrony atop similar resource-constrained hardware. Nevertheless, our work "stands on the shoulders of giants", leveraging decades of work on dependable distributed systems that we revisit in a new context. We provide due account of such literature in Section 4.7, together with a brief description of CPS protocols that provide communication guarantees in specific scenarios.

4.1 System Model

We target typical CPS applications where processing occurs in distinct and periodic sense-process-actuate cycles [SLMR05]. Sensing occurs at nodes equipped with application-specific sensing devices, which periodically report sensed data to nodes with attached actuators. These nodes process the data and drive the actuators accordingly. Unlike mainstream systems, a distinction therefore exists between sensor nodes—which generate data and act as *senders*—and actuator nodes—which consume data and act as *receivers*. Our VIRTUS description is based on such a distinction, although a node may simultaneously act as both.

We accept that both nodes and links between nodes may fail, although such failures do not occur infinitely often or liveness may be compromised. Nodes fail according to a *crash-stop* model [CT96], that is, nodes execute correctly until they *silently* halt and execute no further action. In principle, nothing prevents us from considering a crash-recovery failure model [CT96], where a process silently halts but then recovers from where it left. CPS devices, however, often lack the stable storage required to log information for recovery [LBL+13]. Nevertheless, a crash-stop model fits the reality of deployed systems, where nodes may fail because of battery depletion and lose the previous state upon rebooting when power is again available.

We consider a synchronous and unreliable communication model. This entails that: *i)* there is a known upper bound on message transmission delays, and *ii)* the communication channel may *silently* lose individual messages. The latter, in particular, matches experimental evidence about the *time-varying* nature of network topologies in low-power wireless, for example, due to interference and obstacles [SDTL10].

We do not consider Byzantine failures, which may affect communication or a node's state in ways different than those stipulated by a protocol's actions. For example, messages are either correctly delivered or not delivered at all—a node never processes corrupted messages. Similarly, a node's state always evolves in ways that map to a feasible execution of a protocol's actions. In general, such Byzantine failures require dedicated solutions that we do not consider in this thesis.

4.2   LWB as the Communication Support

We choose to use LWB as the foundation for Virtus, mainly because its bus-like operation eases the design of the interactions required to implement virtual synchrony.   LWB already provides some of the mechanisms required for virtual synchrony, such as:

- Implicit total ordering *if* data *messages are received*, due to the exclusive access to the bus during data slots.

- Explicit join operation *for senders*, in the form of stream requests (see Section 3.1), along with mechanisms to detect possible failures afterwards (see Section 3.2.5).

Moreover, we showed in Section 3.7 that LWB is significantly more efficient than alternative multicast protocols for low-power wireless networks, which makes it practical for resource-constrained devices to bear the overhead required to provide virtual synchrony guarantees.

Nevertheless, using LWB, the gap to provide virtual synchrony includes functionality such as:

- *Guaranteed delivery.* Although LWB delivers messages with high probability, it does not ensure by itself that data messages eventually reach the intended receivers.

- *Total ordering in the presence of communication failures.* The ordering feature in LWB, which is a side-effect of the time-triggered operation, breaks if data messages are not delivered.

- *Explicit join operations for receivers.* These operations, together with mechanisms required to detect receivers' failures, are necessary to create the group.

- *View management.* A notion of view, along with its management as senders and receivers join or fail, is absent in LWB.

We describe next the Virtus functionality that fills this gap.

4.3   Building Up to Virtual Synchrony

Many variants of virtual synchrony exist [CKV01]. We consider the most traditional incarnation, corresponding to the intuitive definition provided in the beginning of this chapter. More formally [Bir05]:

> Given any two nodes P and Q, any two messages ⟨1⟩ and ⟨2⟩ generated in any arbitrary relative order, and any two *consecutive* different views V and V' that include P and Q, we wish to ensure that if P delivers message ⟨1⟩ before message ⟨2⟩ in view V, then Q also delivers ⟨1⟩ before ⟨2⟩ in V.

VIRTUS achieves the above by implementing two core functionality:

- An *atomic multicast service*, providing reliable and totally-ordered multicast delivery at member nodes, illustrated in Section 4.3.2.

- A *view management service*, used at any non-faulty member to maintain the list of view members, described in Section 4.3.3.

We conclude in Section 4.3.4 by proving that our design provides virtual synchrony guarantees.

4.3.1   Overview

VIRTUS provides applications with traditional virtual synchrony operations such as sending and receiving messages, and notification of view changes. Moreover, the application may use a join operation to notify VIRTUS that it intends to join a view as a sender or a receiver. Once being a view member, a node may request at any time to be removed from the view. This provides programmers with a simple API that exposes traditional virtual synchrony operations.

During operation, we distinguish between three disjoints sets of nodes:

- *View members*: non-faulty nodes that are members of the current view V.

- *Participating nodes*: nodes not yet in view V that use (or used) join to notify their intent to join the view (either as senders or receivers).

- *Non-participating nodes*: nodes that only help propagate packets across multiple hops, and thus operate transparently with respect to virtual synchrony.

VIRTUS round. Figure 4.1 shows the communication slots within a VIRTUS round r. It highlights the two types of slots that VIRTUS adds to LWB:

Figure 4.1: Communication slots within a VIRTUS round r. Highlighted in red are the slots added to LWB (cf. Figure 3.2).

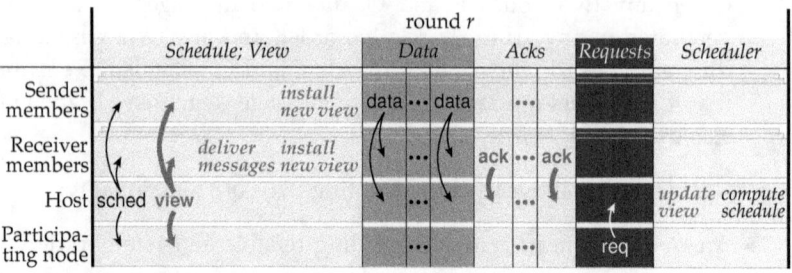

Figure 4.2: Operation and exchange of messages during a VIRTUS round r. Highlighted in red is the functionality added to LWB.

view and ack. The former are used by the host to distribute the current view to all other nodes, the latter are used by receivers to inform the host about the content of their buffers. [1]

Figure 4.2 depicts the VIRTUS operation during round r, and in particular how messages are exchanged among nodes. The VIRTUS-specific processing occurs mainly at five distinct stages:

- *Schedule; View.* After a sched message, the host distributes a view message with the current view:

$$V = \{V.id, V.S, V.R\} \qquad (4.1)$$

  This consists of an identifier $V.id$ and a list of member nodes, split between senders $V.S$ and receivers $V.R$. Based on the received sched and view messages, receivers possibly deliver previously buffered data messages to the application right after processing the view message. Should the received view V differ from the currently installed one, members of the new view perform a *view installation* and deliver a view_change notification to the application.

- *Data.* As in the original LWB, senders in $V.S$ transmit data messages during data slots according to the content of the sched message.

---

[1]To simplify the description of VIRTUS and the analysis of the virtual synchrony guarantees it provides, in the remainder of this chapter we ignore sched messages transmitted by the host at the end of a round (e.g., sched $r + 1$ at the end of round r in Figure 4.1). As described in Section 3.2.4, such schedule transmission is a performance optimization that is not essential for the functioning of LWB and thus of VIRTUS.

Unlike LWB, receivers in $V\!\mathcal{R}$ locally buffer received data messages and wait until the next view message before possibly delivering them to the application.

- *Acks.* After the exchange of data messages, each receiver in $V\!\mathcal{R}$ sends an ack message to inform the host of the set of data messages currently in its buffer. This information is mainly used for reliable delivery in atomic multicast, but also to ensure correct view changes, as we describe next.

- *Requests.* As in LWB, the round ends with a non-allocated req slot, where participating nodes compete to inform the host about their intention to join the view. Different from LWB, participating nodes may be either senders or receivers.

- *Scheduler.* If the host detected the crash of a node or received a request during the req slot, it updates the current view, which it distributes in the next view message. The host then computes the next round schedule, based also on received ack messages and possible view updates. [1]

Concepts and notations. We say that a view member *executes* during a round only if it receives both sched and view messages. Should instead a member fail to receive either of them, it refrains from any processing during the round. The sched message is needed for the original LWB operation, the view message is required to check the current group membership. We also call a round r *stable* if the host receives ack messages in round r from all non-faulty members in $V\!\mathcal{R}$.

For simplicity, the following description considers the host as a non-participating node, although nothing prevents it from being a view member. The discussion does not consider host failures, as they are dealt with by the original LWB failover mechanisms described in Section 3.2.6. We show in Section 4.3.4 that these mechanisms do not break the virtual synchrony guarantees VIRTUS provides.

We express the VIRTUS processing as operations on sets of message identifiers. A data message identifier I_i is a triple $I_i = \{\texttt{sender\_id}, \texttt{stream\_id}, \texttt{generation\_time}\}$ that *uniquely* specifies a data message $\langle i \rangle$ generated by a sender. Table 4.1 summarizes the sets we introduce throughout the chapter.

4.3.2 Atomic Multicast

The atomic multicast service in VIRTUS provides reliable and totally-ordered multicast to view members.

Symbol	Meaning
K_r	Messages scheduled in round r
F_r	Messages scheduled in round r for the first time
C_r	Messages generated by senders expelled from a view in round r
S_r	Messages generated by senders in the view installed at round r
A_r	Messages acknowledged by all non-faulty receivers in round r
B_r^R	Messages in the buffer of receiver R after the data slots of round r
D_r^R	Messages delivered by receiver R during round r
E_r^R	Messages discarded by receiver R during round r

Table 4.1: Sets of data message identifiers used to describe Virtus operation.

4.3.2.1   Virtus Mechanisms for Atomic Multicast

As discussed in Section 4.2, LWB provides neither guaranteed delivery nor total ordering in the presence of communication failures. Virtus employs the following mechanisms to fill the gap:

- Receivers buffer received data messages and use ack messages to inform the host of the data messages in their buffers.

- Based on the received ack messages, the host reallocates slots for data messages missing from at least one receiver buffer.

- After receiving a new sched message, receivers deliver buffered data messages for which no slot is reallocated.

The distributed operation of LWB, where all nodes are potential receiver of all messages, allows to achieve this functionality in a way that is practical for resource-constrained CPS devices. We first explain these mechanisms based on an example where nodes do not fail. We discuss in Section 4.3.3 how to account for the output of the view management service.

4.3.2.2   Example

Figures 4.3–4.6 show several rounds of an example execution in a network with four nodes: one sender S, two receivers P and Q, and a host H. Nodes S, P, and Q are view members and have view $V = \{1, \{S\}, \{P, Q\}\}$ installed. At every round r, sender S has a new data message ready to transmit, $\langle r \rangle$, which is unambiguously specified by data message identifier I_r. For each slot, the figures show messages exchanged, the content of the receiver buffers, and the data messages that P and Q deliver to the application. As no other node intends to join, no req messages are transmitted during these rounds. Because there are also no node failures, view V never changes. At the beginning of round $r = 1$, the receiver buffers are empty.

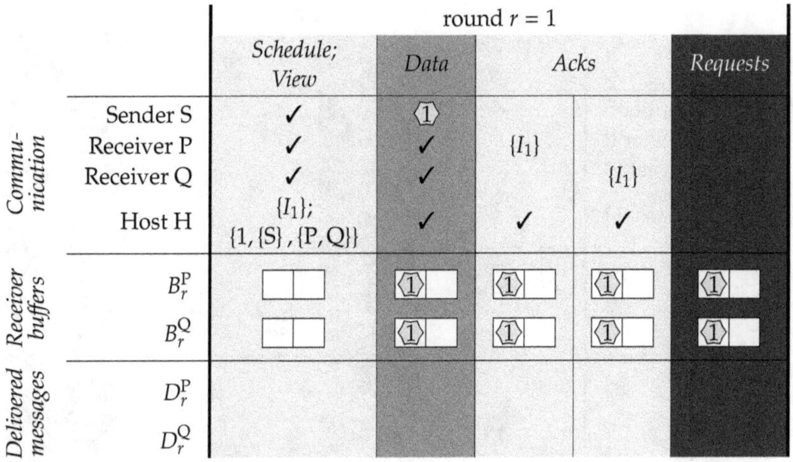

		Schedule; View	Data	Acks	Requests
				round $r = 1$	
Communication	Sender S	✓	①		
	Receiver P	✓	✓	$\{I_1\}$	
	Receiver Q	✓	✓	$\{I_1\}$	
	Host H	$\{I_1\}$; $\{1,\{S\},\{P,Q\}\}$	✓	✓	✓
Receiver buffers	B_r^P	[ ][ ]	⟨①⟩[ ]	⟨①⟩[ ]	⟨①⟩[ ]	⟨①⟩[ ]
	B_r^Q	[ ][ ]	⟨①⟩[ ]	⟨①⟩[ ]	⟨①⟩[ ]	⟨①⟩[ ]
Delivered messages	D_r^P				
	D_r^Q				

Figure 4.3: Example execution of atomic multicast: round $r = 1$. Symbol ✓ denotes a successful reception.

Round $r = 1$ (Figure 4.3). The round starts with the host transmitting schedule $K_1 = \{I_1\}$, which instructs sender S that it can transmit data message ① in the assigned data slot. In general, the schedule K_r for round r is an ordered set of message identifiers $\{I_i, I_j, \dots\}$ that senders can transmit during round r; we omit the additional information in sched messages related to the LWB operation (e.g., the round period T).

All nodes receive the schedule and communicate during the data slot: S transmits message ①; both P and Q receive ① and insert it into their buffers. To ensure total order, receivers buffer multiple messages in the relative order they appear in the sched message. In the two subsequent ack slots, receivers P and Q inform the host about the content of their buffers. Because $B_1^P = B_1^Q = \{I_1\}$, both ack messages include the identifier I_1 of data message ①.

Round $r = 1$ is a stable round, as the host receives ack messages from all receivers in $V.\mathcal{R}$. The host computes the set of data messages it can stop scheduling as $A_1 = \{I_1\}$. In general, for a stable round r, the set of data message identifiers A_r not to reschedule in subsequent rounds is the intersection of the data messages in the receiver buffers B_r^R of any receiver R in $V.\mathcal{R}$:

$$A_r = \bigcap_R B_r^R, \quad \forall R \in V.\mathcal{R}, \quad r \text{ stable} \qquad (4.2)$$

The data messages in A_r are indeed already in the receivers' buffers and do not need to be retransmitted.

		round $r = 2$		
		Schedule; View	Data	Acks	Requests
Commu- nication	Sender S	✓	②		
	Receiver P	✗			
	Receiver Q	✓	✓		$\{I_2\}$
	Host H	$\{I_2\}$; $\{1, \{S\}, \{P, Q\}\}$	✓		✓
Receiver buffers	B_r^P	［① ］	［① ］	［① ］	［① ］	［① ］
	B_r^Q	［ ］	［② ］	［② ］	［② ］	［② ］
Delivered messages	D_r^P				
	D_r^Q	①			

Figure 4.4: Example execution of atomic multicast: round $r = 2$. Symbol ✓ denotes a successful reception; symbol ✗ denotes a communication failure.

Round $r = 2$ (Figure 4.4). The new schedule transmitted by the host specifies that there is only one data slot, for message ②: because all receivers in $V.\mathcal{R}$ received data message ①, the host allocates no more slots for it. The schedule K_r of a generic round r is indeed obtained from the schedule of the previous round K_{r-1} by: *i)* removing the identifiers of data messages acknowledged by all receivers in the previous round, included in A_{r-1}; and *ii)* possibly adding identifiers of newly-generated data messages never scheduled before, included in F_r:

$$K_r = (K_{r-1} \setminus A_{r-1}) \cup F_r \tag{4.3}$$

Figure 4.4 shows that P fails to receive the sched message, thus it does not execute in round $r = 2$ and the content of its buffer does not change: $B_2^P = \{I_1\}$. Differently, Q receives the schedule and from $K_2 = \{I_2\}$ it infers that data message ① reached all receivers in $V.\mathcal{R}$ and can be delivered to the application. We indicate the delivery with $D_2^Q = \{I_1\}$. In general, during a round r, a receiver R delivers data messages that are in its buffer from the previous round B_{r-1}^R and whose identifiers are not included in the current schedule K_r, meaning they reached all receivers:

$$D_r^R = B_{r-1}^R \setminus K_r \tag{4.4}$$

To provide totally-ordered delivery, this operation occurs in the order the data messages are found in the buffer.

During the subsequent data slot, S transmits message ②, which is buffered at Q only, as P is not participating in round $r = 2$ because of

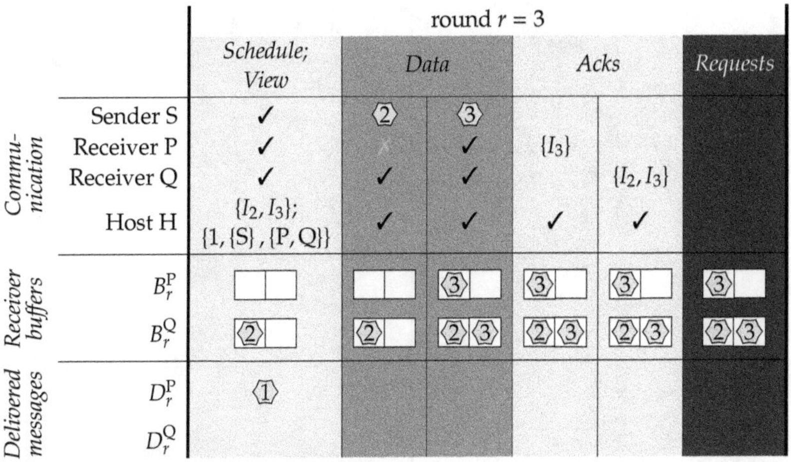

		Schedule; View	Data		Acks		Requests
Communication	Sender S	✓	②	③			
	Receiver P	✓	✗	✓	$\{I_3\}$		
	Receiver Q	✓	✓	✓		$\{I_2, I_3\}$	
	Host H	$\{I_2, I_3\}$; $\{1, \{S\}, \{P, Q\}\}$	✓	✓	✓	✓	
Receiver buffers	B_r^P	☐☐	☐☐	③☐	③☐	③☐	③☐
	B_r^Q	②☐	②☐	②③	②③	②③	②③
Delivered messages	D_r^P	①					
	D_r^Q						

Figure 4.5: Example execution of atomic multicast: round $r = 3$. Symbol ✓ denotes a successful reception; symbol ✗ denotes a communication failure.

the missed schedule. The host receives an ack message from Q but not from P, thus round $r = 2$ is non-stable. The host computes the set of data messages not to reschedule as $A_2 = \varnothing$. This applies for any non-stable round r, as at least one receiver may be missing at least one data message:

$$A_r = \varnothing, \quad r \text{ non-stable} \tag{4.5}$$

Round $r = 3$ (Figure 4.5). For $A_2 = \varnothing$, the schedule reassigns a slot for data message ② in addition to a slot for the new data message ③. From (4.3) indeed follows $K_3 = (\{I_2\} \setminus \varnothing) \cup \{I_3\} = \{I_2, I_3\}$.

This time, both receivers P and Q obtain the schedule. Finally, P realizes that ① reached all receivers because no slots are allocated to it in K_3, and accordingly delivers such data message: with $B_2^P = \{I_1\}$, from (4.4) follows $D_3^P = \{I_1\} \setminus \{I_2, I_3\} = \{I_1\}$. Differently, Q delivers no data messages at this round, because although it already received message ②, a slot is still scheduled for it; with $B_2^Q = \{I_2\}$, from (4.4) indeed follows $D_3^Q = \{I_2\} \setminus \{I_2, I_3\} = \varnothing$.

Based on schedule K_3, sender S retransmits data message ②. P does not receive it, while Q does but immediately drops it as the same message ② is already buffered from the previous round. Message ③ is instead received and buffered by both receivers. The host receives both ack messages $B_3^P = \{I_3\}$ and $B_3^Q = \{I_2, I_3\}$, thus the round is stable. From (4.2) it computes $A_3 = \{I_3\} \cap \{I_2, I_3\} = \{I_3\}$: only message ③ is indeed in both buffers.

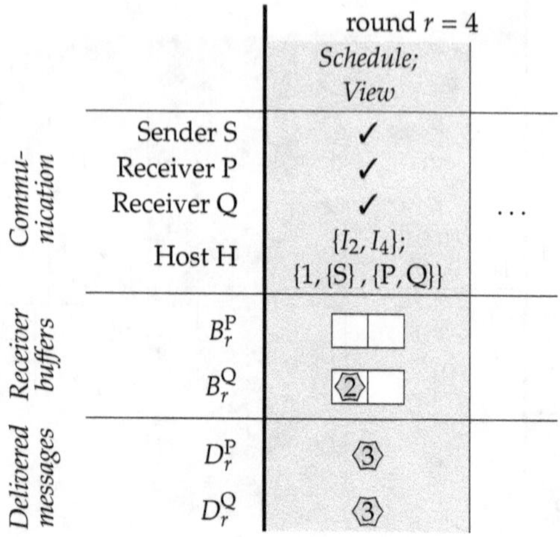

		round $r = 4$ Schedule; View
Communication	Sender S	✓
	Receiver P	✓
	Receiver Q	✓
	Host H	$\{I_2, I_4\}$; $\{1, \{S\}, \{P, Q\}\}$
Receiver buffers	B_r^P	▢▢
	B_r^Q	②▢
Delivered messages	D_r^P	③
	D_r^Q	③

Figure 4.6: Example execution of atomic multicast: beginning of round $r = 4$. Symbol ✓ denotes a successful reception.

Beginning of round $r = 4$ (Figure 4.6). Schedule K_4 specifies that a data slot is again rescheduled for message ②, which P has not received yet, plus another data slot is scheduled for the new message ④: according to (4.3), $K_4 = (\{I_2, I_3\} \setminus \{I_3\}) \cup \{I_4\} = \{I_2, I_4\}$. The reception of this sched message makes both receivers deliver data message ③: from (4.4), $D_4^P = \{I_3\} \setminus \{I_2, I_4\} = \{I_3\}$ and $D_4^Q = \{I_2, I_3\} \setminus \{I_2, I_4\} = \{I_3\}$.

Summary. Throughout the rounds in the example, and in the presence of arbitrary communication failures, receivers P and Q deliver the same data messages ① and ③ in the same order. The key to this functionality is in (4.2)–(4.5). These equations, however, require modifications to account for node crashes and corresponding view changes, as we illustrate next.

4.3.3 View Changes

The view management service informs the application about the current view V, which it updates in response to node crashes or recoveries.

4.3.3.1 Virtus Mechanisms for View Management

As discussed in Section 4.2, LWB has no notion of view and provides no explicit support for receivers. The following mechanisms fill this gap:

- Both senders and receivers compete during req slots in order to join the view. At each round, the host distributes the current view V, possibly updated based on node crashes and received requests.

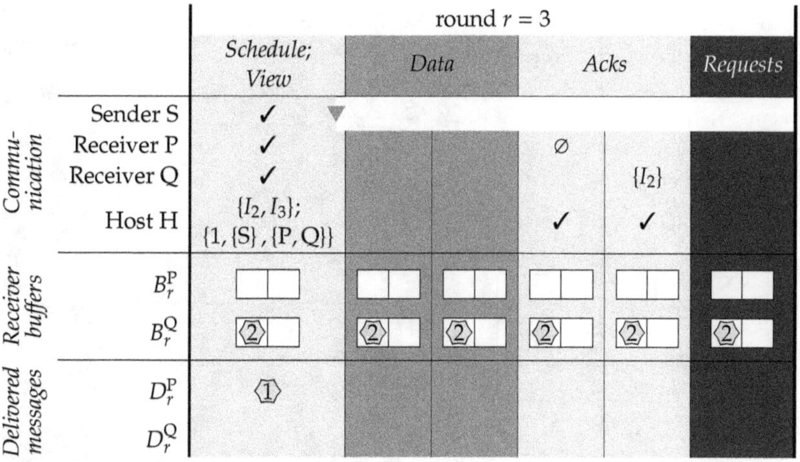

		round $r = 3$					
		Schedule; View	Data		Acks		Requests
Communication	Sender S	✓	▼				
	Receiver P	✓			∅		
	Receiver Q	✓				$\{I_2\}$	
	Host H	$\{I_2, I_3\}$; $\{1, \{S\}, \{P, Q\}\}$			✓	✓	
Receiver buffers	B_r^P	☐☐	☐☐	☐☐	☐☐	☐☐	☐☐
	B_r^Q	⟨2⟩☐	⟨2⟩	⟨2⟩	⟨2⟩	⟨2⟩	⟨2⟩
Delivered messages	D_r^P	⟨1⟩					
	D_r^Q						

Figure 4.7: Example execution with view changes: round $r = 3$. Symbol ✓ denotes a successful reception; symbol ▼ denotes a node crash.

- The host overhears messages exchanged among view members to monitor their continuing operation. Similar to LWB, the host uses a counter-based scheme that marks a view member as crashed when not heard for more than X consecutive rounds, X being a protocol parameter whose tuning we investigate in Section 4.6. Different from LWB, the host monitors not only data messages from senders in $V.S$ but also ack messages from receivers in $V.\mathcal{R}$.

- To provide atomic multicast also in the presence of sender and receiver crashes, delivery occurs only at receivers that are listed in $V.\mathcal{R}$ and only for data messages from senders that belong to $V.S$.

We also observe that the failure detector we use may be inaccurate [CT96] and mistake message loss for node crashes. However, we show that even in case of false positives VIRTUS keeps providing virtual synchrony guarantees. We again use a concrete example to explain how these mechanisms blend together within a realistic VIRTUS execution.

4.3.3.2 Example

We now consider the example execution in Figures 4.7–4.10. The overall setting and the first two rounds are as in Figures 4.3–4.4. For simplicity, we set the number of rounds for detecting node crashes to $X = 1$. This time, the execution unfolds as follows.

Round $r = 3$ (Figure 4.7). As seen before, the schedule for this round is $K_3 = \{I_2, I_3\}$: both receivers obtain it, and P delivers data message ⟨1⟩.

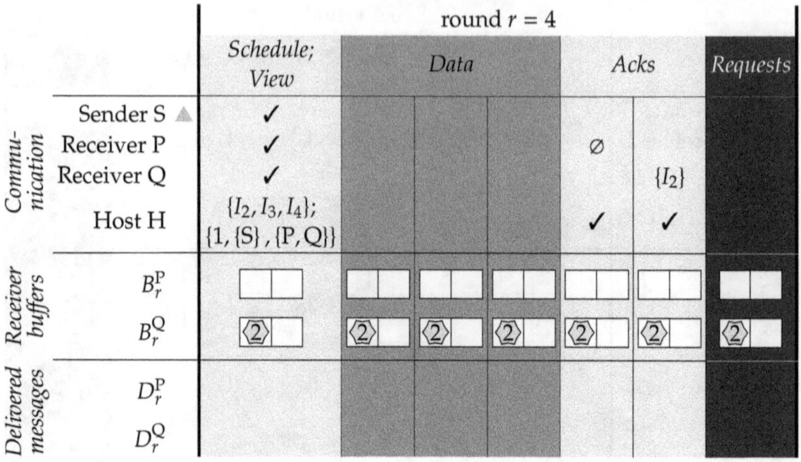

		Schedule; View	Data	Acks	Requests
Communication	Sender S ▲	✓			
	Receiver P	✓		∅	
	Receiver Q	✓		$\{I_2\}$	
	Host H	$\{I_2, I_3, I_4\}$; $\{1, \{S\}, \{P, Q\}\}$		✓ ✓	
Receiver buffers	B_r^P				
	B_r^Q	⟨2⟩	⟨2⟩ ⟨2⟩ ⟨2⟩	⟨2⟩ ⟨2⟩	⟨2⟩
Delivered messages	D_r^P				
	D_r^Q				

Figure 4.8: Example execution with view changes: round $r = 4$. Symbol ✓ denotes a successful reception; symbol ▲ denotes a node recovery.

This time, however, sender S crashes immediately after the view slot. As a result, S cannot (re)transmit data messages as instructed by the schedule. This potentially creates a situation violating atomic multicast: data message ⟨2⟩, already in the buffer of receiver Q, needs to be delivered by both P and Q or neither. As we show next, for simplicity we make the latter happen, based on the crash-stop model we consider for nodes.

In the remainder of the round, the data slots remain unused, thus the buffers at both receivers remain unchanged. Receivers send ack messages $B_3^P = \varnothing$ and $B_3^Q = \{I_2\}$. The host receives both ack messages, thus the round is stable, and according to (4.2) it computes $A_3 = \varnothing$.

Round $r = 4$ (Figure 4.8). With $K_3 = \{I_2, I_3\}$, $A_3 = \varnothing$, and $F_4 = \{I_4\}$, from (4.3) follows schedule $K_4 = \{I_2, I_3, I_4\}$. Sender S recovers before the beginning of this round and executes join. As we consider a crash-stop model, S cannot replay the execution before the crash and transmit data messages ⟨2⟩, ⟨3⟩, ⟨4⟩ according to the schedule. Therefore, we must treat these situations as if the recovered node were a new device, and force the view change corresponding to the crash of the now-recovered node. To accomplish this, the recovered node keeps silent while it sees itself listed in the current view, meaning that the crash was not yet detected and no corresponding view change occurred.

As a result, although sender S receives both sched and view messages, it transmits no data message in round $r = 4$, as it finds itself already listed in the view. Being $X = 1$ in this example, at the end of this round the host detects the crash of S and expels it from the updated view $V = \{2, \varnothing, \{P, Q\}\}$.

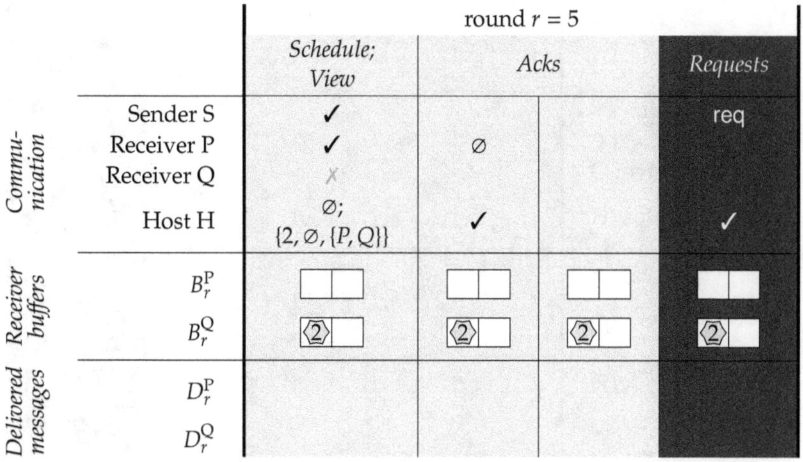

		round $r = 5$		
		Schedule; View	Acks		Requests
Communication	Sender S	✓			req
	Receiver P	✓	∅		
	Receiver Q	✗			
	Host H	∅; $\{2, \varnothing, \{P, Q\}\}$	✓		✓
Receiver buffers	B_r^P	☐☐	☐☐	☐☐	☐☐
	B_r^Q	②☐	②☐	②☐	②☐
Delivered messages	D_r^P				
	D_r^Q				

Figure 4.9: Example execution with view changes: round $r = 5$. Symbol ✓ denotes a successful reception; symbol ✗ denotes a communication failure.

Round $r = 5$ (Figure 4.9). As $V.S = \varnothing$, there is no sender in the view and $F_5 = \varnothing$. However, with $K_4 = \{I_2, I_3, I_4\}$ and $A_4 = \varnothing$, computing the schedule based on (4.3) would incorrectly lead to $K_5 = \{I_2, I_3, I_4\}$. These messages indeed belong to an execution of S that will never be replayed, and S will never retransmit them. Therefore, the host must stop scheduling **data** slots for messages generated by the crashed S. In general, we achieve this by modifying (4.3) as:

$$K_r = [K_{r-1} \setminus (A_{r-1} \cup C_{r-1})] \cup F_r \tag{4.6}$$

Set $C_{r-1}(\subseteq K_{r-1})$ includes the identifiers of **data** messages from crashed senders that have been removed from the view at the end of round $r-1$. In our example, this set corresponds to $C_4 = \{I_2, I_3, I_4\}$. According to (4.6), the schedule for round $r = 5$ is empty: $K_5 = [\{I_2, I_3, I_4\} \setminus (\varnothing \cup \{I_2, I_3, I_4\})] \cup \varnothing = \varnothing$.

During the remainder of this round, receiver P executes and installs the new view. Receiver Q, which still has **data** message ② in its buffer, misses either the **sched** or **view** message, so it does not execute, and is stuck at the previous view that still includes sender S. The now-recovered S finally sees itself not listed in the new view, so it sends a request to join during the **req** slot, and the host receives this **req** message.

Round $r = 6$ (Figure 4.10). Two issues may arise. First, if sender S is immediately readmitted and the view updated again, the new view $\{3, \{S\}, \{P, Q\}\}$ would trick receiver Q to think that S never crashed. Besides the identifier, this view is indeed identical to $\{1, \{S\}, \{P, Q\}\}$, which Q still has installed as it did not execute in round $r = 5$. Second, we need

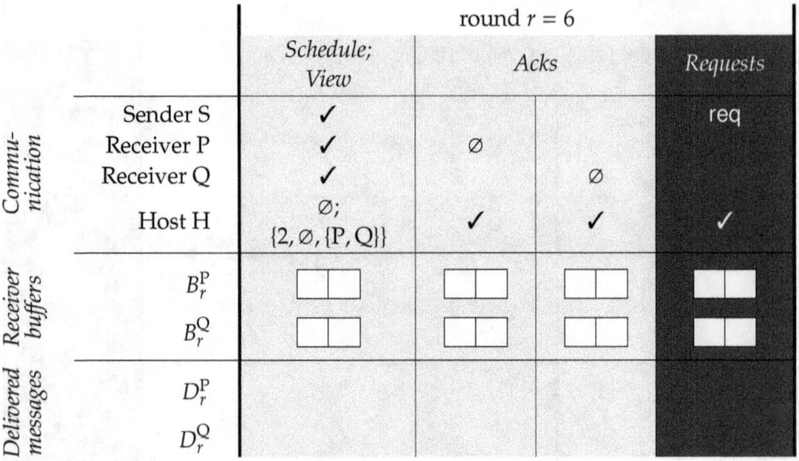

		Schedule; View	Acks		Requests
Commu-nication	Sender S	✓			req
	Receiver P	✓	∅		
	Receiver Q	✓		∅	
	Host H	∅; $\{2, \emptyset, \{P, Q\}\}$	✓	✓	✓
Receiver buffers	B_r^P				
	B_r^Q				
Delivered messages	D_r^P				
	D_r^Q				

Figure 4.10: Example execution with view changes: round $r = 6$. Symbol ✓ denotes a successful reception.

receiver Q to discard data message ② when it installs view $\{2, \emptyset, \{P, Q\}\}$ and realizes that S crashed. As $K_6 = \emptyset$, based on (4.4) receiver Q would deliver $D_6^Q = \{I_2\} \setminus \emptyset = \{I_2\}$. This would violate virtual synchrony, as the other non-faulty receiver P will never deliver ②.

To address these issues, we both postpone adding new senders until after a stable round—ensuring that all non-faulty receivers have the latest view installed—and modify (4.4) as:

$$D_r^R = \left[B_{r-1}^R \setminus K_r \right] \cap S_r \tag{4.7}$$

Set S_r identifies *any* data message generated by senders that are members of the current view. Note that S_r is merely a formal artifact: receivers do *not* need to know the list of messages ever generated by senders. It suffices to check whether a data message that a receiver is about to deliver is generated by a sender that is member of the current view. If so, the data message is delivered. If not, this data message is surely not in S_r, and is discarded as it is not guaranteed that all non-faulty receivers have it in their buffer and are thus able to deliver it.

The set E_r^R of data messages discarded by a receiver R is:

$$E_r^R = \left[B_{r-1}^R \setminus K_r \right] \setminus S_r \tag{4.8}$$

In our example, $E_6^Q = I_2$ and data message ② is discarded at Q because sender S is not member of the current view. Sender S, on the other hand, not seeing itself admitted to the view, transmits the req message also during the req slot at the end of this round.

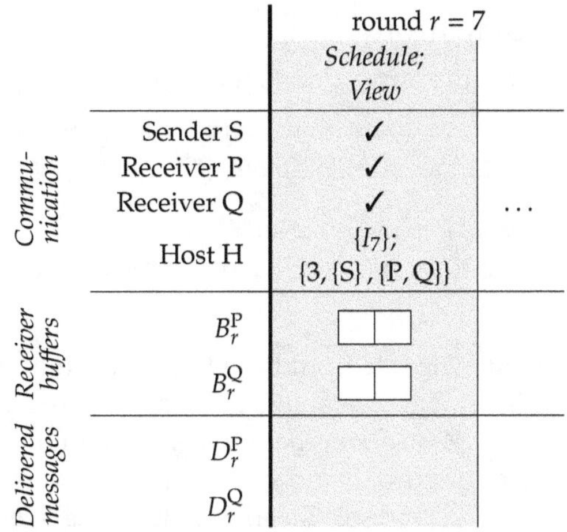

		round $r = 7$	
		Schedule; *View*	
Commu- nication	Sender S	✓	
	Receiver P	✓	
	Receiver Q	✓	...
	Host H	$\{I_7\}$; $\{3, \{S\}, \{P, Q\}\}$	
Receiver buffers	B_r^P	☐☐	
	B_r^Q	☐☐	
Delivered messages	D_r^P		
	D_r^Q		

Figure 4.11: Example execution with view changes: beginning of round $r = 7$. Symbol ✓ denotes a successful reception.

Beginning of round $r = 7$ (Figure 4.11). Round $r = 6$ was stable, so the host finally admits sender S to the view, a view change occurs, and the new view is disseminated to the nodes. The processing resumes normally.

Summary. The example shows how VIRTUS retains atomic multicast also against sender crashes. The processing for receiver crashes is simpler: they can be removed from a view as soon as the crash is detected, and admitted to a view as soon as they send a req. If the host expels a non-faulty receiver from a view due to the repeated loss of ack messages, such receiver empties its buffer before sending a req, as no virtual synchrony guarantees can be provided for data messages already in its buffer.

As described above, we integrate view management with atomic multicast by taking additional care in scheduling and delivering data messages, as reflected in (4.6) and (4.7), and by possibly discarding them, as specified by (4.8). From these equations, and (4.2) and (4.5) from Section 4.3.2, we can also observe that VIRTUS satisfies basic properties of group communication systems [CKV01]:

- *Self inclusion*: a node is a member of a view it installs.

- *Local monotonicity:* identifiers of views installed by a node are monotonically increasing.

- *Initial view event:* delivery of data messages occurs within a view.

- *Primary component membership:* views installed by nodes are totally ordered.

4.3.4   Virtual Synchrony

We now prove that Virtus does guarantee virtual synchrony.

Bounded buffer. First, we show that Virtus determines an upper bound on the number of data messages buffered at a receiver.

Lemma 1. *At the end of a round r, the set B_r^R of message identifiers buffered at a receiver R is a subset of the schedule K_q received by R in the last round $q \leq r$ where R executes.*

Proof. After receiving schedule K_q and view V_q during round q, every non-faulty receiver R in $V_q.\mathcal{R}$ delivers and discards buffered data messages according to (4.7) and (4.8). From that moment and until the next round where R executes, R buffers only messages with identifiers in K_q and that are not already buffered. Indeed, R can add data messages to the buffer only in round q, as it does not execute in any following round.   □

In general, the number of message identifiers that can fit a sched message is bounded by D, for example, due to platform-dependent restrictions on packet sizes (see Section 3.3.1). The cardinality of any schedule K_r is thus bound by D: $|K_r| \leq D$. From Lemma 1, it immediately follows that a receiver has at most D messages buffered at any point in time, and thus:

Theorem 1. *A buffer size of at least D ensures that no buffer overflows occur at a receiver.*

Virtual synchrony. We first prove the following lemma, which we use later to study the virtual synchrony properties provided by Virtus.

Lemma 2. *Every receiver that executes in a stable round r' and is a view member until the next stable round r'' delivers the same set of data messages from the end of r' to the end of r''.*

Proof. According to (4.6), the schedule transmitted by the host at the beginning of round $r' + 1$ is $K_{r'+1} = [K_{r'} \setminus (A_{r'} \cup C_{r'})] \cup F_{r'+1}$. The schedule for the remaining rounds $r = \{r' + 2, \ldots, r''\}$ is $K_r = K_{r-1} \cup F_r$, as the host removes message identifiers from the schedule only *after* a stable round, and only the last round r'' in this sequence is stable; thus, $A_{r-1} = C_{r-1} = \varnothing$ for all rounds $r = \{r' + 2, \ldots, r''\}$. As a result, the schedule transmitted during rounds $r = \{r' + 1, \ldots, r''\}$ is:

$$K_r = [K_{r'} \setminus (A_{r'} \cup C_{r'})] \cup (F_{r'+1} \cup \cdots \cup F_r) \qquad (4.9)$$

Consider the generic receiver R in Figure 4.12, which is a member of every view V_r in rounds $r = \{r', \ldots, r''\}$. Receiver R thus executes in r' and

	⟨1⟩			⟨2⟩			⟨3⟩	
	$r = r'$		$r' < r < \rho$		$r = \rho$		$\rho < r < r''$		$r = r''$
	Schedule; View	Acks	Schedule; View	Acks	Schedule; View	Acks	Schedule; View	Acks	Schedule; View	Acks
R	✓		✗		✓		—		✓	
H	$K_{r'}$; $V_{r'}$	✓	K_r; V_r		K_ρ; V_ρ		K_r; V_r	—	$K_{r''}$; $V_{r''}$	✓
D_r^R	∅		∅		$A_{r'}$		∅		∅	

Figure 4.12: Visual representation for the proof of Lemma 2: any receiver R that is a view member between two stable rounds r' and r'' delivers the same set of data messages $A_{r'}$ between the end of r' and the end of r''. Symbol ✓ denotes a successful reception; symbol ✗ denotes a communication failure; symbol — denotes that the outcome of a reception attempt is irrelevant for the proof.

does not crash between the end of r' and the end of r''. Being round r' stable, the host receives an ack message also from receiver R during this round, which in turn ensures that R receives schedule $K_{r'}$ and has view $V_{r'}$ installed during r'.

Let us name $\rho \in \{r' + 1, \ldots, r''\}$ the first round after r' where R executes. Round ρ is stable if and only if $\rho = r''$. The data messages that receiver R delivers before, during, and after ρ are as follows. Figure 4.12 provides a visual representation of these deliveries.

⟨1⟩ *Rounds* $r = \{r' + 1, \ldots, \rho - 1\}$. Receiver R misses either the sched or the view message, or both, thus it does not execute and delivers no data messages:

$$D_r^R = \emptyset, \quad r = \{r' + 1, \ldots, \rho - 1\} \qquad (4.10)$$

The set of data messages in its buffer remains the one of the last stable round: $B_r^R = B_{r'}^R$.

⟨2⟩ *Round* ρ. Receiver R receives schedule K_ρ and view V_ρ. It delivers data messages as per (4.7), in the same relative order their identifiers appear in the sched message of the last round where R executed, which is r'. From (4.7) and (4.9) descends:

$$D_\rho^R = A_{r'} \qquad (4.11)$$

See Appendix 4.A for the derivation of (4.11). Similarly, as per (4.8) receiver R discards data messages $F_\rho^R = B_{r'}^R \cap C_{r'}$ generated by crashed senders. If view V_ρ differs from the currently installed view $V_{r'}$, R installs V_ρ after the message delivery.

⟨3⟩ *Rounds* $r = \{\rho + 1, \ldots, r''\}$.   If receiver R misses the sched or
the view message, it does not execute and delivers no data
messages; otherwise, it executes and delivers data messages as
per (4.7). However, this set is also empty, because all data messages
acknowledged by all receivers were in $A_{r'}$ and were delivered in ρ,
and no stable round occurs after r' and until r''. Assuming that q is
the last round before r where R executed ($\rho \leq q < r$), we indeed have
$B_{r-1}^{R} \subseteq K_q$ from Lemma 1 and $K_q \subseteq K_r$ from (4.9). The latter is because
the host removes no message identifiers from the schedule between
the end of two consecutive stable rounds r' and r''. By combining
these two properties, $B_{r-1}^{R} \subseteq K_r$, and from (4.7) follows $D_r^{R} = \emptyset$. We
can finally write:

$$D_r^{R} = \emptyset, \quad r = \{\rho + 1, \ldots, r''\} \tag{4.12}$$

Because $A_{r'}$ in (4.11) depends neither on the specific receiver R nor on
the round ρ, every non-faulty receiver delivers the same set of messages
between the end of r' and the end of r''.   □

The example in Figures 4.3–4.5 showed a concrete case between rounds
$r' = 1$ and $r'' = 3$. Despite message loss, both receivers P and Q deliver the
same data message ⟨1⟩. Receiver Q delivers it in round $\rho^Q = 2$, whereas
receiver P delivers it in $\rho^P = 3$. The virtual synchrony property is a direct
consequence of Lemma 2, and is expressed by the following theorem:

Theorem 2. *If two receivers both install the same new view V following the
same previous view V', then they deliver the same set of data messages in V'.*

Proof. Lemma 2 ensures that, while members of the same view
$V_{r'} = \cdots = V_{r''}$, all receivers deliver the same set of messages within that
view.   Moreover, any receiver that installs a new view V_ρ following
view $V_{r'}$ delivers the same set of messages $A_{r'}$ right before installing
the new view.   □

Same view delivery. From Lemma 2 it also follows that:

Theorem 3. *If two receivers deliver the same data message, then they deliver it
in the same view.*

Proof. Based on the proof of Lemma 2, a receiver R delivers data
messages $A_{r'}$ during round ρ and within view $V_{r'}$, the latter being the
view it has installed at the beginning of ρ. As $V_{r'}$ depends neither on the
specific receiver R nor on the round ρ, any receiver that delivers these
messages delivers them within the same view $V_{r'}$.   □

Total ordering. The following theorem ensures that receiver members deliver data messages in the same order.

Theorem 4. *When receivers deliver data messages, they deliver them in the same order.*

Proof. Based on the proof of Lemma 2, a receiver R delivers data messages A_r during round ρ, and with the same relative order their identifiers had in schedule K_r. As this order depends neither on the specific receiver R nor on the round ρ, any receiver member that delivers these messages delivers them in the same order. $\square$

Host failures. We observe that the theorems above hold also in the face of host failures. After a crash of the current host, nodes stop receiving sched or view messages and the *entire* VIRTUS processing stops. If the host does not recover within a specified amount of time, the LWB failover policy elects a different node as the new host (see Section 3.2.6). In this case, the system restarts from scratch, with the new host distributing empty sched and view messages. Senders and receivers thus realize they are not listed in the view and, after discarding all buffered messages, transmit req messages to join the view.

This simple operation entails a performance overhead due to a new bootstrapping process, but it ensures that none of the virtual synchrony guarantees discussed above are violated. Nevertheless, we note that it is possible to employ more complex mechanisms that keep the overhead to a minimum, for example, by making the new host reuse information included in the last view it received from the crashed host.

4.4   FIFO Delivery

Fault-tolerant distributed systems often require messages to be delivered in the same order they are generated [Sch90, CKV01]. In addition to total ordering, VIRTUS provides per-node and system wide FIFO delivery by means of very limited modifications.

Per-node FIFO ordering. The default scheduling policy of LWB, which we inherit also in VIRTUS, ensures that the host schedules in FIFO order slots for data messages from each sender (see Section 3.3.2). In VIRTUS, however, non-faulty receivers may violate the per-node FIFO ordering when delivering data messages, due to retransmissions. This happens, for example, in Figures 4.3–4.5: receivers P and Q deliver message ③ before message ②. This is because, during a round r, the per-node FIFO ordering holds only within the set of messages scheduled for the first time F_r. The ordering of delivery, however, depends also on when data messages are acknowledged by all receivers.

We can provide per-node FIFO delivery with a simple modification to the scheduling algorithm at the host. The key idea is to keep scheduling slots for data messages even though these are already acknowledged by all non-faulty receivers, should these messages be generated later than non-acknowledged data messages from the same sender. In the example of Figures 4.3–4.5, this entails rescheduling data slots for message ⟨3⟩ in round $r = 4$, even if it was already acknowledged by both P and Q. Because the identifier of message ⟨3⟩ keeps appearing in the schedule, both receivers do not deliver it as per (4.7).

Specifically, when the host computes schedule K_{r+1} at the end of a stable round r, in (4.6) it uses $A_r^{nF} \subseteq A_r$ instead of A_r. The set A_r^{nF} includes data messages in A_r whose generation time at the sender is not greater than the generation time of any other message in K_r from *the same sender*.

System-wide FIFO ordering.   Similarly, system-wide FIFO delivery entails that receivers deliver no data messages generated before already delivered messages, regardless of the sender.

This requires two modifications. First, we change the LWB scheduler such that the system-wide FIFO ordering holds within data messages scheduled for the first time, included in F_r. This is possible because the host knows when each sender generates new data messages, due to their periodic traffic patterns. Second, similarly to the per-node FIFO delivery above, we modify how the host decides which data messages to reschedule. Specifically, when the host computes schedule K_{r+1} at the end of a stable round r, in (4.6) it uses $A_r^{sF} \subseteq A_r$ instead of A_r. The set $A_r^{sF} \subseteq A_r$ includes data messages in A_r whose generation time is not greater than the generation time of any other message in K_r, *regardless of the sender*.

With either of these modifications, Virtus maintains totally-ordered delivery, because the mechanisms at the receivers remain the same. FIFO delivery, however, entails allocating slots not strictly needed, as the corresponding data messages are already buffered at all non-faulty receivers, introducing additional overhead. We evaluate the impact of this overhead in Section 4.6.

4.5   Implementation

We implement Virtus on top of the Contiki operating system [Conb, DGV04], targeting the TelosB platform [PSC05]. The mechanisms added to LWB occupy only 6 kB of program memory: considering that our LWB implementation requires 22 kB (see Table 3.2), Virtus occupies in total

28 kB of program memory, leaving 20 kB available for the application.

Compared to the original LWB implementation, we reduce from 60 to 40 the maximum number of data slots D allocated per round. This makes our prototype support up to 15 ack slots and thus up to 15 receivers, but it decreases the bandwidth available for data messages. This setting is representative of existing CPS deployments where virtual synchrony may be necessary [SLMR05, CCD+11, KLS+10, BvKH+11, Sch12]. Nevertheless, designers can tune this value based on application requirements. All other functional parameters retain the original LWB values shown in Table 3.1.

To reduce energy consumption, the host allocates ack slots only when needed, that is, in rounds with at least one data slot or between a view update and the next stable round. The latter is to ensure that all receivers install an updated view, as discussed in Section 4.3.3. Finally, we apply an optimization to overcome the loss of view messages. If a node detects that the current view V—whose identifier $V.id$ is embedded also in the sched message—is the same as the one it has already installed, it executes even if it misses the view message, as it already knows V from view messages it received in previous rounds.

4.6 Evaluation

VIRTUS incurs a runtime overhead compared with LWB's best-effort operation. We use results from testbed experiments to quantitatively assess this aspect, and to study the impact of different parameter settings.

4.6.1 Testbeds and Metrics

Testbeds. We evaluate VIRTUS using two sensor network testbeds. TWIST is an indoor installation of 90 TelosB nodes spanning three floors of a university building [HKWW06]. On TWIST we use an intermediate transmit power of -15 dBm, yielding a network diameter of 4 hops. FLOCKLAB includes 30 TelosB nodes at a university building, and has a network diameter of 4 hops at the maximum transmit power of 0 dBm [LFZ+13b]; four of these nodes are located outdoors.

To factor out sources of network unreliability we cannot control, we use channel 26 to minimize interference with co-located Wi-Fi networks. We artificially emulate message loss during ad-hoc experiments. In all experiments, data messages carry a payload of 15 bytes. Unless otherwise stated, we set $X = 10$ as the threshold to detect node crashes; we further discuss this choice in Section 4.6.4. The specific traffic profile varies depending on the type of experiment.

Metrics. To assess the performance overhead in exchange for virtual synchrony, we consider an unmodified LWB as the baseline. For both LWB and VIRTUS, we measure:

- *Data latency*, defined as the interval from when the application at a sender sends a data message to when a receiver delivers that message to the application.

- *Radio duty cycle*, defined as the fraction of time a node has the radio turned on, commonly regarded as an indication of energy efficiency in low-power wireless protocols [AY05].

We expect virtual synchrony to impact both: latency should increase because messages are delivered only after *all* receivers buffered them and notified the host about it; radio duty cycle should increase because of additional control traffic and retransmissions absent in LWB.

Complementary to these figures, we assess how effective are the virtual synchrony guarantees VIRTUS provides by measuring:

- *Per-receiver data yield*, defined as the fraction of generated data messages successfully delivered at each receivers.

- *System-wide data yield*, defined as the fraction of generated data messages successfully delivered at *all* receivers.

- *View latency*, defined as the interval from when a view member crashes to when all non-faulty nodes install an updated view.

In the absence of crashes, per-receiver and system-wide data yield should measure 100 % due to atomic multicast. Nevertheless, it is interesting to check the LWB performance in the same settings, to relate the gap to virtual synchrony with the performance overhead. View latency is instead useful to understand the trade-offs for different parameter settings.

4.6.2   Cost of Virtual Synchrony

Scenario. On TWIST, we randomly pick 45 senders and let them generate one data message per minute, addressed to a variable number of 2, 5, 10, and 15 receivers across different experiments. The remaining nodes only help propagate packets in the network. These settings depict scenarios akin to typical CPS deployments [SLMR05, CCD+11, KLS+10, BvKH+11, Sch12]. For each scenario, we run 1-hour long experiments with LWB and VIRTUS, the latter with different delivery policies: no FIFO, per-node FIFO, and system-wide FIFO. We fix the round period to $T = 10\,$s. No failures are artificially injected in the network.

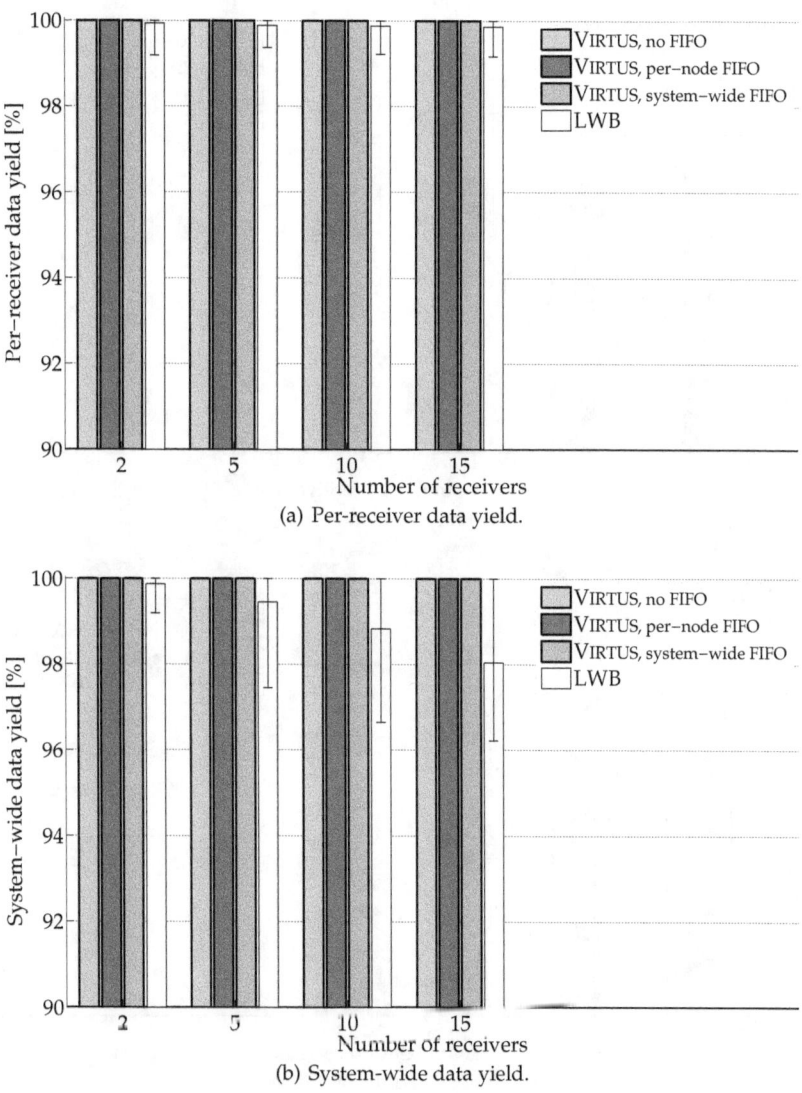

(a) Per-receiver data yield.

(b) System-wide data yield.

Figure 4.13: Atomic multicast on TWIST, for different numbers of receivers and types of ordered delivery. Bars denote averages; error bars indicate 5th and 95th percentiles.

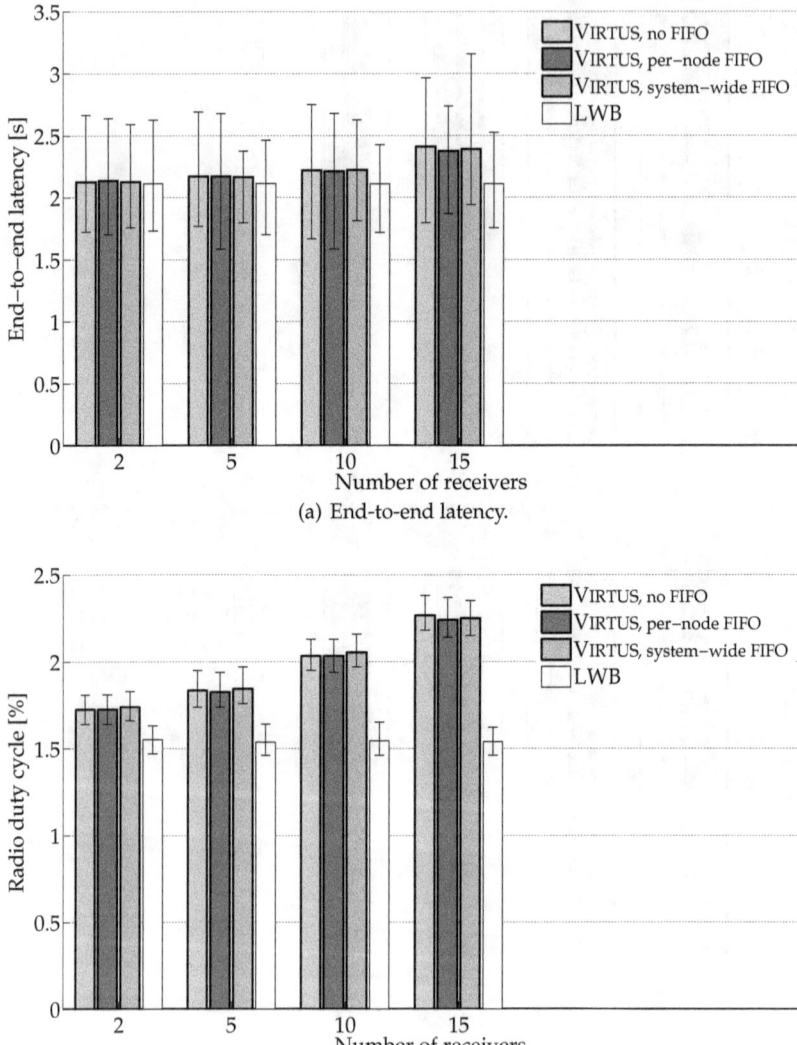

(a) End-to-end latency.

(b) Radio duty cycle.

Figure 4.14:   Cost of virtual synchrony on TWIST, for different numbers of receivers and types of ordered delivery. Bars denote averages; error bars indicate 5th and 95th percentiles.

Results. We first verify that atomic multicast in VIRTUS delivers all data messages, regardless of the type of ordered delivery. Figure 4.13 indeed shows that VIRTUS achieves 100 % per-receiver and system-wide data yield across the board. We also see in Figure 4.13(a) that LWB's per-receiver data yield is largely independent of the number of receivers: each receiver delivers on average about 99.89 % of data messages. With LWB, however, the fraction of data messages that are delivered by *all* receivers decreases with the number of receivers, as shown in Figure 4.13(b). With two receivers, for example, 99.87 % of data messages are delivered by both receivers; with 15 receivers, system-wide data yield averages only 98.04 %. This is because with more receivers it is more likely that *at least one* of them misses a certain data message.

Figure 4.14 plots the performance overhead in data latency and radio duty cycle. Figure 4.14(a) shows that with two receivers LWB and VIRTUS deliver messages with similar average latency of 2.11 s and 2.13 s, respectively. As LWB's reliability already approaches 100 %, most data messages indeed require no retransmissions and the processing in VIRTUS resembles that in LWB. With more receivers, it is more likely that at least one data or ack message is lost. In VIRTUS, this triggers retransmissions from the senders and buffering at the receivers to provide atomic multicast according to Section 4.3.2. This, however, results only in a slight increase in latency, which averages 2.39 s with 15 receivers. The type of ordered delivery has little impact on latency: because of few retransmissions, only in rare cases the host reallocates data slots for already acknowledged messages to enforce FIFO delivery as described in Section 4.4.

Figure 4.14(b) shows the energy overhead of VIRTUS compared with LWB. Two aspects contribute to this: the additional control traffic due to view and ack messages, and the retransmissions of data messages in case of losses. With two receivers, the average radio duty cycle in VIRTUS is only 0.18 % higher than in LWB: 1.73 % against 1.55 %. More receivers entail more ack slots and a higher probability that data or ack messages are lost. For these reasons, different from LWB, the radio duty cycle with VIRTUS increases with the number of receivers. Nevertheless, its average is less than 2.25 % even with 15 receivers. The energy overhead is again largely independent of the type of ordered delivery. Notably, this figure is way smaller than in most existing best-effort multicast protocols for low-power wireless. For instance, on the same TWIST testbed and in a similar scenario with 8 receivers and 45 senders generating one data message per minute, Muster + LPL requires an average radio duty cycle of 11.54 % to deliver only 98.67 % of data messages (see Figure 3.13).

Next, we show how these trade-offs evolve when the system operates with highly unreliable wireless communication.

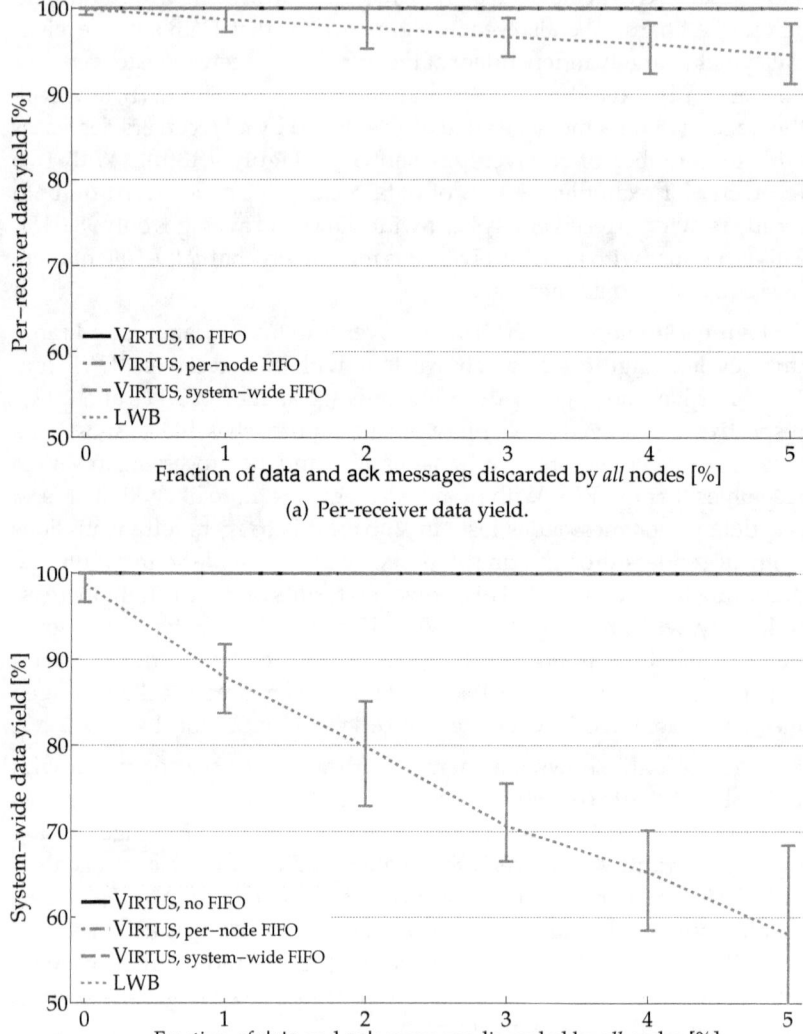

(a) Per-receiver data yield.

(b) System-wide data yield.

Figure 4.15: Atomic multicast on TWIST when communication failures are artificially injected, for different types of ordered delivery. Lines denote averages; error bars indicate 5th and 95th percentiles.

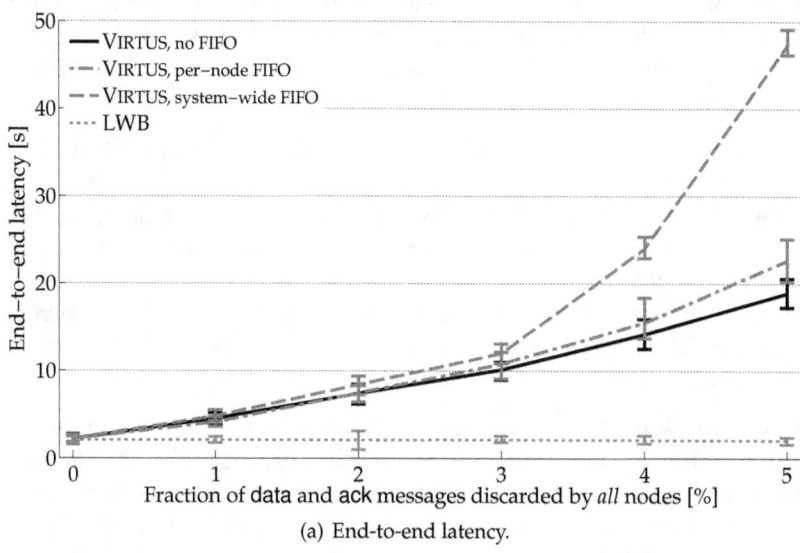

(a) End-to-end latency.

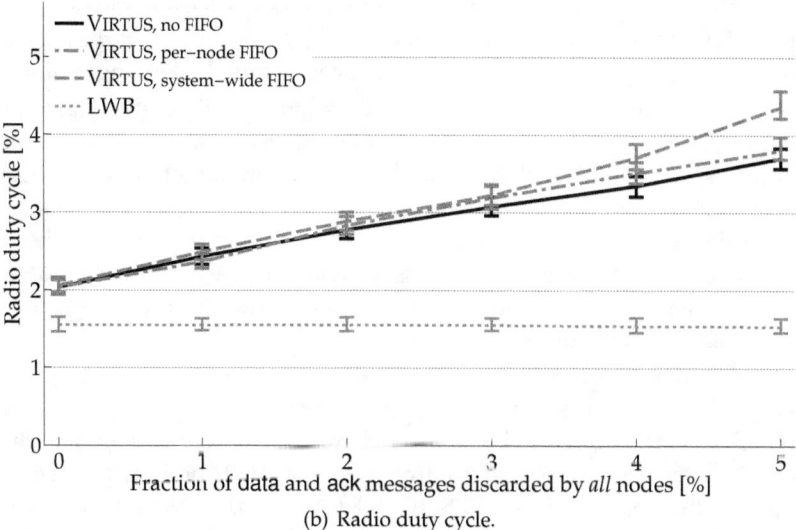

(b) Radio duty cycle.

Figure 4.16: Performance on TWIST when communication failures are artificially injected, for different types of ordered delivery. Lines denote averages; error bars indicate 5th and 95th percentiles.

4.6.3   Resilience to Network Unreliability

Cyber-physical systems are often employed in scenarios with significant network unreliability, for example, due to external wireless interference [LPLT10]. We evaluate the resilience of VIRTUS to these scenarios by injecting artificial message loss.

Scenario. On TWIST, we use 45 nodes as senders and 10 as receivers. We make *all 90 nodes* randomly discard between 1 % and 5 % of messages in data and ack slots, in 1 % steps. Similar scenarios are very challenging; say every node in a 4-hop route drops 5 % of messages: a simple best-effort protocol would yield only about 81 % of the messages *at a single receiver*. Similar settings are extremely unlikely to occur in real deployments. Nevertheless, they are useful to understand the behavior of VIRTUS with respect to network reliability. All other settings are as in Section 4.6.2.

Results. Figure 4.15 shows that LWB's per-receiver and system-wide data yield increasingly suffer as the network becomes less reliable, due to its best-effort operation and lack of retransmissions. When every node discards 5 % of data messages, for example, each receiver delivers on average 94.54 % of data messages, but only 58 % of data messages are delivered by all 10 receivers. Atomic multicast in VIRTUS instead provides 100 % per-receiver and system-wide data yield across the board. In addition, it guarantees total ordering even in this challenging setting.

The cost for this performance is illustrated in Figure 4.16(a) and Figure 4.16(b), showing data latency and radio duty cycle as the network becomes less reliable. With LWB, both metrics are largely independent of network reliability: only one slot is allocated for each data message regardless of how many receivers successfully receive it. In VIRTUS, these figures increase as the network is less reliable, because more slots for data and ack messages are allocated—possibly across multiple rounds—before the receivers finally deliver.

Particularly, Figure 4.16(a) shows that data latency grows significantly. However, these values are within tolerance of most CPS applications, whose dynamics often follow slowly-changing environmental phenomena (e.g., temperature), and control loops run with periods of several minutes [SLMR05, CCD+11, KLS+10, BvKH+11, Sch12]. Nevertheless, we note that this high data latency in VIRTUS is also due to the fact that the round period is set to $T = 10$ s in these experiments. By using smaller values for the round period T, and possibly adapting it at runtime as in Section 3.3.1, we can further reduce data latency at the cost of increased radio duty cycle. As for radio duty cycle in Figure 4.16(b), the performance in absolute terms is again better than many multicast protocols for low-power wireless that only provide best-effort operation [AY05].

By comparing different ordering policies in Figure 4.16, one notes that the performance loss is higher when enforcing FIFO delivery, and in particular system-wide FIFO. Indeed, the latter entails the allocation of data slots for a message until *all* previously-generated messages have been acknowledged by *all* receivers, which in turn delays message delivery.

4.6.4 Influence of Parameter Setting

A key parameter in VIRTUS is X: the number of rounds the host must not hear from a view member to detect a crash. We now show that this value significantly impacts the resulting view latency.

An illustrative example. We analyze a 1-hour run on FLOCKLAB with four senders S1, S2, S3, and S4 generating one data message every 10 s, and four receivers R1, R2, R3, and R4. We emulate node crashes and recoveries by making them not communicate (i.e., crash) at a random instant and then reboot (i.e., recover) with a random delay between 0 s and 600 s after the crash. The round period T is 2 s; X is set to 10.

Figure 4.17(a) shows a 11-minute excerpt of the VIRTUS operation. Nodes start with view 8 installed, and receivers deliver four data messages every 10 s. At $t = 57.4$ s receiver R1 crashes; the non-faulty nodes install view 9, whose set of receivers does not include R1, at $t = 82$ s, yielding a view latency of 24.6 s. Figure 4.17(b) zooms into this time interval and shows the contribution of X to this view latency.

⟨1⟩ From $t = 57.4$ s to $t = 60$ s: the host allocates no data or ack slots at $t = 58$ s because the senders have no new messages to transmit. Because of this, it can not detect the crash of R1. The length of this stage thus depends on the interleaving of the node crash and existing traffic. Indeed, the host can not detect the crash of a sender or a receiver when no data or ack traffic is supposed to occur.

⟨2⟩ From $t = 60$ s to $t = 80$ s: the host allocates data and ack slots for newly generated messages at $t = 60$ s, 62 s, . . . , but it receives no ack messages from R1. Active receivers R2, R3, and R4 buffer received data messages. The length of this stage is $X \times T$: 20 s in this example.

⟨3⟩ From $t = 80$ s to $t = 82$ s: the host detects at the end of the round starting at $t = 80$ s that it received no ack messages from R1 for more than $X = 10$ rounds (✖ in the figure). As a result, it updates the view by removing R1 and starts distributing it in the next round. Non-faulty members receive and install the new view already at $t = 82$ s. Active receivers can finally deliver buffered data messages. The length of this stage thus depends on the round period T and on when nodes successfully receive the new view.

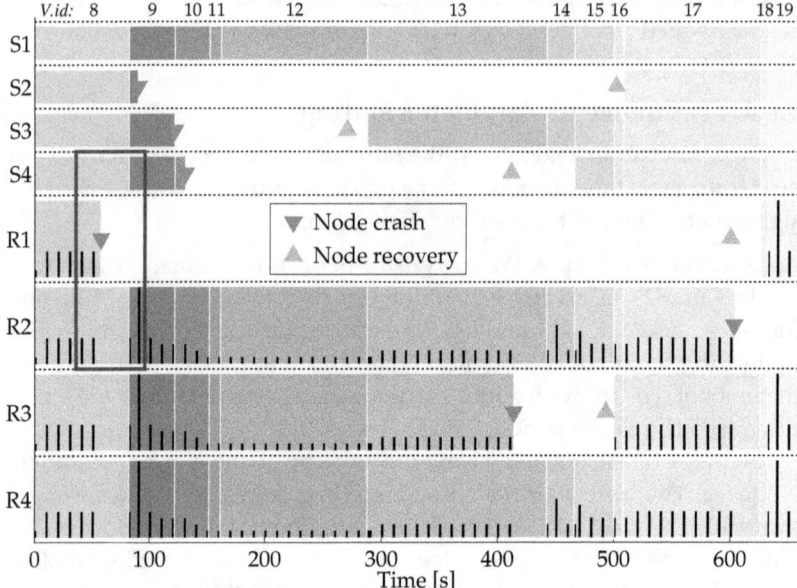

(a) Views installed (colored backgrounds, and corresponding identifiers on top) and goodput at the receivers in data messages delivered per round (black bars) when four senders and four receivers randomly crash and recover.

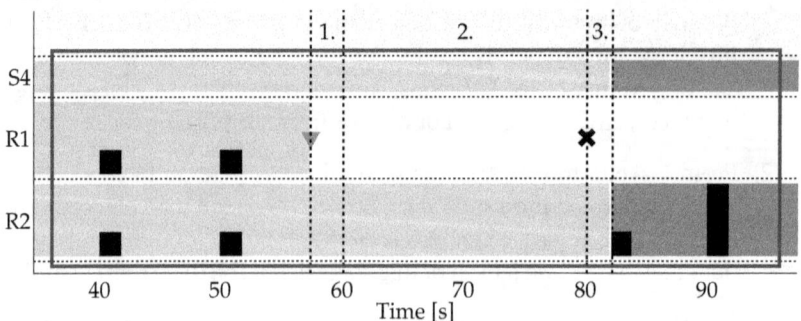

(b) Breakdown of view latency after the crash of a receiver.

Figure 4.17: Virtus operation across view changes. Symbol ▼ denotes a node crash; symbol ▲ denotes a node recovery.

X	0% communication failures, 25 node crashes		10% communication failures, 0 node crashes	
	View latency	*Radio duty cycle*	*False positives*	*Radio duty cycle*
1	20.64 s	2.18 %	59	3.42 %
2	21.58 s	2.22 %	8	3.22 %
3	21.80 s	2.27 %	1	3.06 %
4	22.49 s	2.27 %	0	3.07 %
10	39.34 s	2.51 %	0	3.04 %
20	56.35 s	2.95 %	0	3.10 %

Table 4.2: Impact of threshold X on VIRTUS performance.

After nodes install view 9, senders S2, S3, and S4 crash one after the other. As shown in Figure 4.17(a), these events cause active receivers to deliver less data messages after each crash. The following view changes occur with latencies between 30 s and 32 s, due to executions similar to the one above. As expected, within each view all active receivers deliver the same amount of messages. This confirms that our VIRTUS prototype satisfies the virtual synchrony properties discussed in Section 4.3.4, as we also empirically verify after every experiment.

Setting parameter X. The example shows that the most critical stage is ⟨2⟩, the one directly affected by parameter X. Finding a suitable value for X entails exploring a critical trade-off. A small X allows the host to rapidly detect node crashes and update the view; if X is too small, however, it may mistake message loss for crashes and trigger unnecessary view changes.

To understand this trade-off, we run 1-hour experiments on FLOCKLAB with 24 senders, generating one data message every 30 s, and 5 receivers. The round period is fixed to $T = 2$ s. In one series of experiments, we emulate the crash of 25 view members and inject no artificial message loss. In another series of experiments, nodes do not crash but discard 10% of data and ack messages. In both cases, we vary X between 1 and 20 across different experiments.

Table 4.2 reports the results. The left columns in the table confirm that a larger value of X causes a higher view latency, because the host awaits more rounds before removing a crashed node from the view. The additional data slots unnecessarily allocated to a crashed node also cause the radio duty cycle to increase with X. However, the right columns show that with severe network unreliability a low value of X may lead to false positives, causing unnecessary view changes. The radio duty cycle also increases when X is too small, as nodes that are wrongly removed from a view need to re-send requests to join during req slots.

The default value $X = 10$ in our prototype is sufficiently high to minimize the probability of false positives, while also providing reasonable view latencies. Nevertheless, a user can fine-tune the value of X according to the application requirements and the foreseeable amount of message loss.

4.7 Related Work

Virtus bridges research efforts in two previously unrelated areas: virtual synchrony and low-power wireless.

Virtual synchrony lies in a vein of research originated from seminal work [PSL80] on distributed agreement. In similar cases, however, the authors often consider a Byzantine environment, a failure model we do not study. Different flavors and implementations of virtual synchrony emerged over the years [BJ87], often to explore the trade-off between provided guarantees and runtime overhead. Admittedly, our incarnation almost corresponds to the "textbook" definition [Bir05], as we hope it serves as a stepping stone for others. The work by Chockler et al. [CKV01], who systematically survey group communication systems, helped us relate our solutions to the existing literature.

In low-power wireless, solutions exist to provide communication guarantees in specific application scenarios. For example, structural health monitoring applications [CMP+09] often require guaranteed message delivery from multiple sensors to a single data sink. Protocols such as RCRT [PG07] and several ad-hoc solutions [CMP+09], for instance, provide such functionality. Different from Virtus, however, these protocols only support a many-to-one traffic pattern. This is a mismatch against the sense-process-actuate cycles of CPS applications, which generally require many-to-many coordination. In addition, these protocols seldom provide any guarantee against node crashes. Low-power multicast protocols [AY05, MP11], on the other hand, typically provide only best-effort operation.

4.8 Summary

This chapter presented Virtus, a virtually-synchronous messaging layer conceived for extremely resource-constrained devices. Virtus provides atomic multicast and view management in CPS applications with a combination of dedicated techniques that build on an existing best-effort communication layer. We formally proved the correctness of such techniques and used extensive real-world experiments to assess their limited performance overhead compared with best-effort operation.

The value of VIRTUS lies in opening to cyber-physical systems a vast and established literature on dependable distributed systems that builds upon virtual synchrony or variations thereof. We maintain that this will increase the dependability of CPS applications to an extent that is not achievable without relying on such sound conceptual basis.

4.A   Delivery between Successive Stable Rounds

We compute here the set of messages delivered by receiver R during round ρ, after it receives schedule K_ρ and view V_ρ. To this end, we first combine (4.7) and (4.9):

$$D_\rho^R = \left[B_{\rho-1}^R \setminus \left[(K_{r'} \setminus (A_{r'} \cup C_{r'})) \cup \left(F_{r'+1} \cup \cdots \cup F_\rho\right)\right]\right] \cap S_\rho$$

Because R does not execute in rounds $r = \{r'+1, \ldots, \rho-1\}$ (see also Figure 4.12), the set of data messages in its buffer between the end of round r' and the end of round $\rho-1$ does not change: $B_{\rho-1}^R = B_{r'}^R$. For the same reason, before round ρ receiver R buffers no data messages scheduled for the first time during rounds $r = \{r'+1, \ldots, \rho\}$, thus $B_{r'}^R \setminus (F_{r'+1} \cup \cdots \cup F_\rho) = \varnothing$. Because the set of active senders does not change between the end of two consecutive stable rounds r' and r'', we also have that $S_\rho = S_{r'+1}$. We can thus rewrite D_ρ^R as:

$$D_\rho^R = \left[B_{r'}^R \setminus [K_{r'} \setminus (A_{r'} \cup C_{r'})]\right] \cap S_{r'+1}$$
$$= \left[\left((A_{r'} \cup C_{r'}) \cap B_{r'}^R\right) \cup \left(B_{r'}^R \setminus K_{r'}\right)\right] \cap S_{r'+1}$$

Based on Lemma 1 and knowing that receiver R executes during stable round r', $B_{r'}^R \subseteq K_{r'}$. As a result, $B_{r'}^R \setminus K_{r'} = \varnothing$, and D_ρ^R reduces to:

$$D_\rho^R = \left[(A_{r'} \cup C_{r'}) \cap B_{r'}^R\right] \cap S_{r'+1}$$
$$= [(A_{r'} \cap S_{r'+1}) \cup (C_{r'} \cap S_{r'+1})] \cap B_{r'}^R$$

However, $A_{r'} \cap S_{r'+1} = A_{r'}$, because all messages in $A_{r'}$ are from senders that are members during round $r'+1$. By contrast, $C_{r'} \cap S_{r'+1} = \varnothing$, because all messages in $C_{r'}$ are from senders that crashed by round r' and that are not members during $r'+1$. As a result, we can further simplify D_ρ^R to:

$$D_\rho^R = A_{r'} \cap B_{r'}^R$$

According to (4.2), for stable round r' we have that $A_{r'} \subseteq B_{r'}^R$, thus we can finally write the result in (4.11):

$$D_\rho^R = A_{r'}$$

5

Conclusions and Outlook

Cyber-physical systems (CPSs) employ embedded computers and networks that monitor and control physical processes, usually with feedback loops where physical phenomena affect computation and vice versa. Examples of potential cyber-physical systems are systems for assisted living to assist and improve the quality of life of seniors living alone, networked building control systems (such as HVAC and lighting) to improve energy efficiency and demand variability, and systems to protect and improve critical infrastructures.

In order to realize such systems, CPS designers need to face several challenges related to the safety-critical nature of most envisioned applications. In particular, designers need to ensure that deployed systems operate dependably yet efficiently. However, existing low-power wireless communication protocols typically operate in a best-effort manner and do not provide guarantees (e.g., on message delivery orderings) necessary to apply to cyber-physical systems established concepts for the design and validation of dependable distributed systems.

5.1 Contributions

In this thesis, we argued that it is possible to enable dependable communication in cyber-physical systems without sacrificing efficiency by employing a wireless bus—a time-triggered communication infrastructure for multi-hop low-power wireless networks similar to a shared bus. To support our argument, we implemented three main building blocks contributing towards a dependable wireless bus.

Glossy. We first designed Glossy, a novel architecture that provides fast and highly reliable one-to-all communication combined with accurate global time synchronization in multi-hop low-power wireless networks. Glossy achieves this by leveraging synchronous transmissions of the same packet and exploiting constructive interference of multiple baseband signals, while not requiring nodes to know the multi-hop network topology. Its innovative design has also served to foster the development of several new communication protocols by independent authors in the sensor network community.

Low-Power Wireless Bus (LWB). We then developed LWB, an efficient and versatile wireless bus. By using only Glossy floods for communication and employing a time-triggered, centralized operation, LWB turns a multi-hop wireless network into an infrastructure similar to a shared bus where all nodes are potential receivers of all data. LWB enables one-to-many, many-to-one, and many-to-many communication with a performance similar or significantly better than state-of-the-art solutions, ensures fair bandwidth allocation, supports traffic loads and network topologies varying over time, retains its efficiency when mobile nodes roam around a static infrastructure.

VIRTUS. Finally, we extended LWB's best-effort operation to implement VIRTUS, a wireless bus that provides virtual synchrony guarantees. By employing an atomic multicast service and view management service, VIRTUS guarantees that the same set of messages are delivered in the same order at all non-faulty nodes of a low-power wireless network. This allows to apply to cyber-physical systems well-established fault tolerance methods based on replication techniques. To the best of our knowledge, VIRTUS is the first protocol that provides formally-proven ordered message delivery atop similarly resource-constrained hardware.

5.2 Possible Future Directions

We maintain that the building blocks implemented in this thesis may provide the stepping stones necessary to bridging the current gap between cyber-physical systems and concepts from the dependable distributed systems literature. Nevertheless, our work can be extended along several directions to broaden the spectrum of CPS applications that can benefit from communication protocols with delivery guarantees. We identify four major scenarios that in our opinion deserve further investigation: real-time systems, heterogenous wireless networks, large-scale networks, and highly-mobile networks.

Real-time systems.   Several cyber-physical systems require timely message delivery [SLMR05]. In these real-time systems, data streams are associated with deadlines (i.e., strict time constraints), and a communication protocol must guarantee that all data are delivered within given deadlines. We believe that the time-triggered operation employed by a wireless bus, which mimics existing architectures for distributed real-time control applications [Kop11], provides the support required to enable real-time communication in cyber-physical systems. Apposite scheduling strategies can also be developed, possibly leveraging existing algorithms for hard real-time computing systems [But11].

Heterogeneous wireless networks.   For all protocols presented in this thesis, we implemented prototypes targeting TelosB devices compliant with the IEEE 802.15.4 standard. However, other types of platforms and wireless communication standards, such as cellphones and Wi-Fi, may be more suitable, for example, for CPS applications where humans are directly involved in the loop. We believe that our building blocks have the potential to enable dependable communication also in heterogeneous wireless networks, provided that the underlying mechanisms leveraged by our protocols are ported to different wireless embedded platforms.

Large-scale networks.   Our solutions can also be extended to support large-scale networks involving hundreds or thousands of nodes. For example, hierarchies of buses could effectively be employed in smart cities where physical phenomena are monitored and controlled in order to optimize public costs or improve the quality of living [NBH+11]. In such scenarios, clusters of closely-located devices can form multiple low-level wireless buses, while several nodes from each cluster can connect also to a high-level wireless bus to exchange data among clusters.

Highly-mobile networks.   Certain CPS applications, such as systems for traffic control and safety, are highly mobile in nature. Although all the protocols presented in this thesis inherently support mobile nodes, their operation is not optimized for networks that continuously partition and merge as a result of mobility. To achieve the desired trade-off between energy efficiency and responsiveness to rapidly-changing network topologies, these types of scenarios require dedicated solutions that we believe can be integrated into our existing protocols.

Bibliography

[ABN08]   B. C. Arnold, N. Balakrishnan, and H. Nagaraja. *A First Course in Order Statistics*. SIAM, 2008.

[ADB+04]   A. Arora, P. Dutta, S. Bapat, V. Kulathumani, H. Zhang, V. Naik, V. Mittal, H. Cao, M. Demirbas, M. Gouda, Y. Choi, T. Herman, S. Kulkarni, U. Arumugam, M. Nesterenko, A. Vora, and M. Miyashita. A line in the sand: A wireless sensor network for target detection, classification, and tracking. *Elsevier Computer Networks*, 46(5), 2004.

[Atm]   Atmel. AT86RF230 datasheet. `http://www.atmel.com/Images/doc5131.pdf`.

[AY05]   K. Akkaya and M. Younis. A survey on routing protocols for wireless sensor networks. *Elsevier Ad Hoc Networks*, 3(3), 2005.

[BEP+06]   J. Burke, D. Estrin, A. Parker, N. Ramanathan, S. Reddy, and M. B. Srivastava. Participatory sensing. In *Proceedings of the 1st Workshop on World-Sensor-Web (WSW)*, 2006.

[BGH+09]   J. Beutel, S. Gruber, A. Hasler, R. Lim, A. Meier, C. Plessl, I. Talzi, L. Thiele, C. Tschudin, M. Woehrle, and M. Yuecel. PermaDAQ: A scientific instrument for precision sensing and data recovery under extreme conditions. In *Proceedings of the 8th ACM/IEEE International Conference on Information Processing in Sensor Networks (IPSN)*, 2009.

[Bir05]   K. P. Birman. *Reliable Distributed Systems: Technologies, Web Services, and Applications*. Springer, 2005.

[BIS+08]   G. Barrenetxea, F. Ingelrest, G. Schaefer, M. Vetterli, O. Couach, and M. Parlange. SensorScope: Out-of-the-box environmental monitoring. In *Proceedings of the 7th ACM/IEEE International Conference on Information Processing in Sensor Networks (IPSN)*, 2008.

[BJ87]   K. P. Birman and T. A. Joseph. Exploiting virtual synchrony in distributed systems. In *Proceedings of the 11th ACM Symposium on Operating Systems Principles (SOSP)*, 1987.

[But11]   G. Buttazzo. *Hard Real-Time Computing Systems: Predictable Scheduling Algorithms and Applications*. Springer, 2011.

[BvKH+11]   D. J. A. Bijwaard, W. A. P. van Kleunen, P. J. M. Havinga, L. Kleiboer, and M. J. J. Bijl. Industry: Using dynamic WSNs in smart logistics for fruits and pharmacy. In *Proceedings of the 9th ACM Conference on Embedded Networked Sensor Systems (SenSys)*, 2011.

[BvRW07]   N. Burri, P. von Rickenbach, and R. Wattenhofer. Dozer: Ultra-low power data gathering in sensor networks. In *Proceedings of the 6th ACM/IEEE International Conference on Information Processing in Sensor Networks (IPSN)*, 2007.

[CCD+11]   M. Ceriotti, M. Corrà, L. D'Orazio, R. Doriguzzi, D. Facchin, Ş. Gunǎ, G. P. Jesi, R. Lo Cigno, L. Mottola, A. L. Murphy, M. Pescalli, G. P. Picco, D. Pregnolato, and C. Torghele. Is there light at the ends of the tunnel? Wireless sensor networks for adaptive lighting in road tunnels. In *Proceedings of the 10th ACM/IEEE International Conference on Information Processing in Sensor Networks (IPSN)*, 2011.

[CCT+13]   D. Carlson, M. Chang, A. Terzis, Y. Chen, and O. Gnawali. Forwarder selection in multi-transmitter networks. In *Proceedings of the 9th IEEE International Conference on Distributed Computing in Sensor Systems (DCOSS)*, 2013.

[Chu87]   J. C.-I. Chuang. The effects of time delay spread on portable radio communications channels with digital modulation. *IEEE Journal on Selected Areas in Communications*, 5(5), 1987.

[CKJL09]   J. I. Choi, M. A. Kazandjieva, M. Jain, and P. Levis. The case for a network protocol isolation layer. In *Proceedings of the 7th ACM Conference on Embedded Networked Sensor Systems (SenSys)*, 2009.

[CKV01]   G. V. Chockler, I. Keidar, and R. Vitenberg. Group communication specifications: A comprehensive study. *ACM Computing Surveys (CSUR)*, 33(4), 2001.

[CLBR10]   O. Chipara, C. Lu, T. C. Bailey, and G.-C. Roman. Reliable clinical monitoring using wireless sensor networks: Experiences in a step-down hospital unit. In *Proceedings of the 8th ACM Conference on Embedded Networked Sensor Systems (SenSys)*, 2010.

[CMP+09]   M. Ceriotti, L. Mottola, G. P. Picco, A. L. Murphy, Ş. Gunǎ, M. Corrà, M. Pozzi, D. Zonta, and P. Zanon. Monitoring heritage buildings with wireless sensor networks: The Torre Aquila deployment. In *Proceedings of the 8th ACM/IEEE International Conference on Information Processing in Sensor Networks (IPSN)*, 2009.

[CONa]   CONET integrated testbed. https://conet.us.es/cms/.

[Conb]   Contiki operating system. http://www.contiki-os.org/.

[CT96]   T. D. Chandra and S. Toueg. Unreliable failure detectors for reliable distributed systems. *Journal of the ACM (JACM)*, 43(2), 1996.

[DBK$^+$07]   M. Dyer, J. Beutel, T. Kalt, P. Oehen, L. Thiele, K. Martin, and P. Blum. Deployment support network: A toolkit for the development of WSNs. In *Proceedings of the 4th European Conference on Wireless Sensor Networks (EWSN)*, 2007.

[DC08]   P. Dutta and D. Culler. Practical asynchronous neighbor discovery and rendezvous for mobile sensing applications. In *Proceedings of the 6th ACM Conference on Embedded Networked Sensor Systems (SenSys)*, 2008.

[DC09]   P. Dutta and D. Culler. Mobility changes everything in low-power wireless sensornets. In *Proceedings of the 12th USENIX Workshop on Hot Topics in Operating Systems (HotOS XII)*, 2009.

[DCL13]   M. Doddavenkatappa, M. C. Chan, and B. Leong. Splash: Fast data dissemination with constructive interference in wireless sensor networks. In *Proceedings of the 10th USENIX Symposium on Networked Systems Design and Implementation (NSDI)*, 2013.

[DDHC$^+$10] P. Dutta, S. Dawson-Haggerty, Y. Chen, C.-J. M. Liang, and A. Terzis. Design and evaluation of a versatile and efficient receiver-initiated link layer for low-power wireless. In *Proceedings of the 8th ACM Conference on Embedded Networked Sensor Systems (SenSys)*, 2010.

[DEM$^+$10] V. Dyo, S. A. Ellwood, D. W. Macdonald, A. Markham, C. Mascolo, B. Pásztor, S. Scellato, N. Trigoni, R. Wohlers, and K. Yousef. Evolution and sustainability of a wildlife monitoring sensor network. In *Proceedings of the 8th ACM Conference on Embedded Networked Sensor Systems (SenSys)*, 2010.

[DG80]   D. H. Davis and S. A. Gronemeyer. Performance of slotted ALOHA random access with delay capture and randomized time of arrival. *IEEE Transactions on Communications*, 28(5), 1980.

[DGV04]   A. Dunkels, B. Grönvall, and T. Voigt. Contiki - A lightweight and flexible operating system for tiny networked sensors. In *Proceedings of the 1st IEEE Workshop on Embedded Networked Sensors (EmNetS-I)*, 2004.

[DIM97]   S. Dolev, A. Israeli, and S. Moran. Uniform dynamic self-stabilizing leader election. *IEEE Transactions on Parallel and Distributed Systems*, 8(4), 1997.

[DMEST08]   P. Dutta, R. Musăloiu-E., I. Stoica, and A. Terzis. Wireless ACK collisions not considered harmful. In *Proceedings of the 7th ACM Workshop on Hot Topics in Networks (HotNets-VII)*, 2008.

[DMT+11]   A. Dunkels, L. Mottola, N. Tsiftes, F. Österlind, J. Eriksson, and N. Finne. The announcement layer: Beacon coordination for the sensornet stack. In *Proceedings of the 8th European Conference on Wireless Sensor Networks (EWSN)*, 2011.

[DSGS09]   A. Dutta, D. Saha, D. Grunwald, and D. Sicker. SMACK - A SMart ACKnowledgment scheme for broadcast messages in wireless networks. In *Proceedings of the ACM SIGCOMM Conference*, 2009.

[EAR+06]   E. Ertin, A. Arora, R. Ramnath, V. Naik, S. Bapat, V. Kulathumani, M. Sridharan, H. Zhang, H. Cao, and M. Nesterenko. Kansei: A testbed for sensing at scale. In *Proceedings of the 5th ACM/IEEE International Conference on Information Processing in Sensor Networks (IPSN)*, 2006.

[ET90]   A. Ephremides and T. V. Truong. Scheduling broadcasts in multihop radio networks. *IEEE Transactions on Communications*, 38(4), 1990.

[FW10]   R. Flury and R. Wattenhofer. Slotted programming for sensor networks. In *Proceedings of the 9th ACM/IEEE International Conference on Information Processing in Sensor Networks (IPSN)*, 2010.

[GFJ+09]   O. Gnawali, R. Fonseca, K. Jamieson, D. Moss, and P. Levis. Collection tree protocol. In *Proceedings of the 7th ACM Conference on Embedded Networked Sensor Systems (SenSys)*, 2009.

[GLJ11]   A. Gonga, O. Landsiedel, and M. Johansson. MobiSense: Power-efficient micro-mobility in wireless sensor networks. In *Proceedings of the 7th IEEE International Conference on Distributed Computing in Sensor Systems (DCOSS)*, 2011.

[HC04]   J. W. Hui and D. Culler. The dynamic behavior of a data dissemination protocol for network programming at scale. In *Proceedings of the 2nd ACM Conference on Embedded Networked Sensor Systems (SenSys)*, 2004.

[HKWW06] V. Handziski, A. Köpke, A. Willig, and A. Wolisz. TWIST: A scalable and reconfigurable testbed for wireless indoor experiments with sensor networks. In *Proceedings of the 2nd ACM International Workshop on Multi-hop Ad Hoc Networks (REALMAN)*, 2006.

[HV13]   K. Hewage and T. Voigt. Poster abstract: Towards TCP communication with the Low-Power Wireless Bus. In *Proceedings of the 11th ACM Conference on Embedded Networked Sensor Systems (SenSys)*, 2013. To appear.

[IEE03]   IEEE Standards Association. Std. 802.15.4-2003. http://standards.ieee.org/getieee802/download/802.15.4-2003.pdf, 2003.

[JCH84]   R. K. Jain, D.-M. W. Chiu, and W. R. Hawe. A quantitative measure of fairness and discrimination for resource allocation in shared computer systems. Technical Report 301, DEC, 1984.

[KDK+89]   H. Kopetz, A. Damm, C. Koza, M. Mulazzani, W. Schwabl, C. Senft, and R. Zainlinger. Distributed fault-tolerant real-time systems: The Mars approach. *IEEE Micro*, 9(1), 1989.

[KG93]   H. Kopetz and G. Grünsteidl. TTP - A time-triggered protocol for fault-tolerant real-time systems. In *Proceedings of the 23rd International Symposium on Fault-Tolerant Computing (FTCS-23)*, 1993.

[KGK07]   S. Katti, S. Gollakota, and D. Katabi.   Embracing wireless interference: Analog network coding. In *Proceedings of the ACM SIGCOMM Conference*, 2007.

[KKP99]   J. M. Kahn, R. H. Katz, and K. S. J. Pister.   Next century challenges: Mobile networking for "smart dust". In *Proceedings of the 5th ACM/IEEE International Conference on Mobile Computing and Networking (MobiCom)*, 1999.

[KLS+10]   J. Ko, C. Lu, M. B. Srivastava, J. A. Stankovic, A. Terzis, and M. Welsh. Wireless sensor networks for healthcare. *Proceedings of the IEEE*, 98(11), 2010.

[KLW+09]   B. Kusy, H. Lee, M. Wicke, N. Milosavljevic, and L. Guibas. Predictive QoS routing to mobile sinks in wireless sensor networks. In *Proceedings of the 8th ACM/IEEE International Conference on Information Processing in Sensor Networks (IPSN)*, 2009.

[Kop11]   H. Kopetz.   *Real-Time Systems: Design Principles for Distributed Embedded Applications*. Springer, 2011.

[KPC+07]   S. Kim, S. Pakzad, D. Culler, J. Demmel, G. Fenves, S. Glaser, and M. Turon. Health monitoring of civil infrastructures using wireless sensor networks. In *Proceedings of the 6th ACM/IEEE International Conference on Information Processing in Sensor Networks (IPSN)*, 2007.

[KPD13]   Y.-S. Kuo, P. Pannuto, and P. Dutta. Demo abstract: Floodcasting, a data dissemination service supporting real-time actuation and control. In *Proceeding of the 11th ACM International Conference on Mobile Systems, Applications, and Services (MobiSys)*, 2013.

[KPSD12]   Y.-S. Kuo, P. Pannuto, T. Schmid, and P. Dutta.   Reconfiguring the software radio to improve power, price, and portability. In *Proceedings of the 10th ACM Conference on Embedded Networked Sensor Systems (SenSys)*, 2012.

[KWL+11]   M. Keller, M. Woehrle, R. Lim, J. Beutel, and L. Thiele. Comparative performance analysis of the PermaDozer protocol in diverse deployments. In *Proceedings of the 6th International Workshop on Practical Issues in Building Sensor Network Applications (SenseApp)*, 2011.

[Lan08]   K. Langendoen. Medium access control in wireless sensor networks. In *Medium Access Control in Wireless Networks*. Nova Science Publishers, 2008.

[LBL+13]   Y. Lee, S. Bang, I. Lee, Y. Kim, G. Kim, M. H. Ghaed, P. Pannuto, P. Dutta, D. Sylvester, and D. Blaauw. A modular 1 mm^3 die-stacked sensing platform with low power I^{2}C inter-die communication and multi-modal energy harvesting. *IEEE Journal of Solid-State Circuits*, 48(1), 2013.

[Lee08]   E. A. Lee. Cyber physical systems: Design challenges. In *Proceedings of the 11th IEEE Symposium on Object Oriented Real-Time Distributed Computing (ISORC)*, 2008.

[Lee09]   E. A. Lee. Computing needs time. *Communications of the ACM (CACM)*, 52(5), 2009.

[LF76]   K. Leentvaar and J. H. Flint. The capture effect in FM receivers. *IEEE Transactions on Communications*, 24(5), 1976.

[LFZ13a]   O. Landsiedel, F. Ferrari, and M. Zimmerling. Capture at scale: Ultra-fast wireless all-to-all communication. In *Proceedings of the 11th ACM Conference on Embedded Networked Sensor Systems (SenSys)*, 2013. To appear.

[LFZ+13b]   R. Lim, F. Ferrari, M. Zimmerling, C. Walser, P. Sommer, and J. Beutel. FlockLab: A testbed for distributed, synchronized tracing and profiling of wireless embedded systems. In *Proceedings of the 12th ACM/IEEE International Conference on Information Processing in Sensor Networks (IPSN)*, 2013.

[LKA+10]   J. W. Lee, B. Kusy, T. Azim, B. Shihada, and P. Levis. Whirlpool routing for mobility. In *Proceedings of the 11th ACM International Symposium on Mobile Ad Hoc Networking and Computing (MobiHoc)*, 2010.

[LL08]   K. Lin and P. Levis. Data discovery and dissemination with DIP. In *Proceedings of the 7th ACM/IEEE International Conference on Information Processing in Sensor Networks (IPSN)*, 2008.

[LLL+09]   C.-J. M. Liang, J. Liu, L. Luo, A. Terzis, and F. Zhao. RACNet: A high-fidelity data center sensing network. In *Proceedings of the 7th ACM Conference on Embedded Networked Sensor Systems (SenSys)*, 2009.

[LPCS04]   P. Levis, N. Patel, D. Culler, and S. Shenker. Trickle: A self-regulating algorithm for code propagation and maintenance in wireless sensor networks. In *Proceedings of the 1st USENIX Symposium on Networked Systems Design and Implementation (NSDI)*, 2004.

[LPLT10]   C.-J. M. Liang, N. B. Priyantha, J. Liu, and A. Terzis. Surviving Wi-Fi interference in low power ZigBee networks. In *Proceedings of the 8th ACM Conference on Embedded Networked Sensor Systems (SenSys)*, 2010.

[LSW09]   C. Lenzen, P. Sommer, and R. Wattenhofer. Optimal clock synchronization in networks. In *Proceedings of the 7th ACM Conference on Embedded Networked Sensor Systems (SenSys)*, 2009.

[LW09]   J. Lu and K. Whitehouse. Flash flooding: Exploiting the capture effect for rapid flooding in wireless sensor networks. In *Proceedings of the 28th IEEE International Conference on Computer Communications (INFOCOM)*, 2009.

[LWHS02]   D. Li, K. D. Wong, Y. H. Hu, and A. M. Sayeed. Detection, classification, and tracking of targets. *IEEE Signal Processing Magazine*, 19(2), 2002.

[Mar04]   M. Maróti. Directed flood-routing framework for wireless sensor networks. In *Proceedings of the 5th ACM/IFIP/USENIX International Conference on Middleware (Middleware)*, 2004.

[MCP+02]   A. Mainwaring, D. Culler, J. Polastre, R. Szewczyk, and J. Anderson. Wireless sensor networks for habitat monitoring. In *Proceedings of the 1st ACM International Workshop on Wireless Sensor Networks and Applications (WSNA)*, 2002.

[MKSL04]   M. Maróti, B. Kusy, G. Simon, and A. Lédeczi. The flooding time synchronization protocol. In *Proceedings of the 2nd ACM Conference on Embedded Networked Sensor Systems (SenSys)*, 2004.

[MP11]   L. Mottola and G. P. Picco. MUSTER: Adaptive energy-aware multi-sink routing in wireless sensor networks. *IEEE Transactions on Mobile Computing*, 10(12), 2011.

[MSKG10]   S. Moeller, A. Sridharan, B. Krishnamachari, and O. Gnawali. Routing without routes: The backpressure collection protocol. In *Proceedings of the 9th ACM/IEEE International Conference on Information Processing in Sensor Networks (IPSN)*, 2010.

[MT06]   R. Makowitz and C. Temple. Flexray - A communication network for automotive control systems. In *Proceedings of the 7th IEEE International Workshop on Factory Communication Systems (WFCS)*, 2006.

[NBH⁺11]   M. Naphade, G. Banavar, C. Harrison, J. Paraszczak, and R. Morris. Smarter cities and their innovation challenges. *IEEE Computer*, 44(6), 2011.

[NTCS99]   S.-Y. Ni, Y.-C. Tseng, Y.-S. Chen, and J.-P. Sheu. The broadcast storm problem in a mobile ad hoc network. In *Proceedings of the 5th ACM/IEEE International Conference on Mobile Computing and Networking (MobiCom)*, 1999.

[PG07]   J. Paek and R. Govindan. RCRT: Rate-controlled reliable transport for wireless sensor networks. In *Proceedings of the 5th ACM Conference on Embedded Networked Sensor Systems (SenSys)*, 2007.

[PGZM12]   D. Puccinelli, S. Giordano, M. Zuniga, and P. J. Marrón. Broadcast-free collection protocol. In *Proceedings of the 10th ACM Conference on Embedded Networked Sensor Systems (SenSys)*, 2012.

[PHC04]   J. Polastre, J. Hill, and D. Culler. Versatile low power media access for wireless sensor networks. In *Proceedings of the 2nd ACM Conference on Embedded Networked Sensor Systems (SenSys)*, 2004.

[PSC05]   J. Polastre, R. Szewczyk, and D. Culler. Telos: Enabling ultra-low power wireless research. In *Proceedings of the 4th ACM/IEEE International Symposium on Information Processing in Sensor Networks (IPSN)*, 2005.

[PSL80]   M. C. Pease, R. E. Shostak, and L. Lamport. Reaching agreement in the presence of faults. *Journal of the ACM (JACM)*, 27(2), 1980.

[PSLN⁺12]   M. Pajic, S. Sundaram, J. Le Ny, G. J. Pappas, and R. Mangharam. Closing the loop: A simple distributed method for control over wireless networks. In *Proceedings of the 11th ACM/IEEE International Conference on Information Processing in Sensor Networks (IPSN)*, 2012.

[RLSS10]   R. Rajkumar, I. Lee, L. Sha, and J. Stankovic. Cyber-physical systems: The next computing revolution. In *Proceedings of the 47th ACM/IEEE Design Automation Conference (DAC)*, 2010.

[Rus01]   J. M. Rushby. Bus architectures for safety-critical embedded systems. In *Proceedings of the 1st International Workshop on Embedded Software (EMSOFT)*, 2001.

[RWAM05]   I. Rhee, A. Warrier, M. Aia, and J. Min. Z-MAC: A hybrid MAC for wireless sensor networks. In *Proceedings of the 3rd ACM Conference on Embedded Networked Sensor Systems (SenSys)*, 2005.

[RWMX06]   I. Rhee, A. Warrier, J. Min, and L. Xu. DRAND: Distributed randomized TDMA scheduling for wireless ad-hoc networks. In *Proceedings of the 7th ACM International Symposium on Mobile Ad Hoc Networking and Computing (MobiHoc)*, 2006.

[RZS+08]   M. Rossi, G. Zanca, L. Stabellini, R. Crepaldi, A. F. Harris, III, and M. Zorzi. SYNAPSE: A network reprogramming protocol for wireless sensor networks using fountain codes. In *Proceedings of the 5th IEEE Conference on Sensor, Mesh and Ad Hoc Communications and Networks (SECON)*, 2008.

[SAM03]   Y. Sankarasubramaniam, I. F. Akyildiz, and S. W. McLaughlin. Energy efficiency based packet size optimization in wireless sensor networks. In *Proceedings of the 1st IEEE International Workshop on Sensor Network Protocols and Applications (SNPA)*, 2003.

[Sch90]   F. B. Schneider. Implementing fault-tolerant services using the state machine approach: A tutorial. *ACM Computing Surveys*, 22(4), 1990.

[Sch09]   T. Schmid. *Time in Wireless Embedded Systems*. PhD thesis, University of California, Los Angeles (UCLA), 2009.

[Sch12]   J. Schlick. Cyber-physical systems in factory automation - Towards the 4th industrial revolution. In *Proceedings of the 9th IEEE International Workshop on Factory Communication Systems (WFCS)*, 2012.

[SDS10]   T. Schmid, P. Dutta, and M. B. Srivastava. High-resolution, low-power time synchronization an oxymoron no more. In *Proceedings of the 9th ACM/IEEE International Conference on Information Processing in Sensor Networks (IPSN)*, 2010.

[SDTL10]   K. Srinivasan, P. Dutta, A. Tavakoli, and P. Levis. An empirical study of low-power wireless. *ACM Transactions on Sensor Networks (TOSN)*, 6(2), 2010.

[SGE06]   T. Schoellhammer, B. Greenstein, and D. Estrin. Hyper: A routing protocol to support mobile users of sensor networks. Technical Report 2013, CENS, 2006.

[SHM+08]   J. Song, S. Han, A. K. Mok, D. Chen, M. Lucas, M. Nixon, and W. Pratt. WirelessHART: Applying wireless technology in real-time industrial process control. In *Proceedings of the 14th IEEE Real-Time and Embedded Technology and Applications Symposium (RTAS)*, 2008.

[SHSM06]   F. Stann, J. Heidemann, R. Shroff, and M. Z. Murtaza. RBP: Robust broadcast propagation in wireless networks. In *Proceedings of the 4th ACM Conference on Embedded Networked Sensor Systems (SenSys)*, 2006.

[SLMR05]   J. A. Stankovic, I. Lee, A. Mok, and R. Rajkumar. Opportunities and obligations for physical computing systems. *Computer*, 38, 2005.

[SMSM06]   B. Sirkeci-Mergen, A. Scaglione, and G. Mergen. Asymptotic analysis of multistage cooperative broadcast in wireless networks. *IEEE Transactions on Information Theory*, 52(6), 2006.

[Sta08]   J. A. Stankovic. When sensor and actuator networks cover the world. *ETRI Journal*, 30(5), 2008.

[SXLC10]   A. Saifullah, Y. Xu, C. Lu, and Y. Chen. Real-time scheduling for WirelessHART networks. In *Proceedings of the 31st IEEE Real-Time Systems Symposium (RTSS)*, 2010.

[SZHT07]   A. Swami, Q. Zhao, Y.-W. Hong, and L. Tong. *Wireless Sensor Networks: Signal Processing and Communications*. Wiley, 2007.

[TC05]   G. Tolle and D. Culler. Design of an application-cooperative management system for wireless sensor networks. In *Proceedings of the 2nd European Workshop on Wireless Sensor Networks (EWSN)*, 2005.

[Texa]   Texas Instruments. CC1101 datasheet. http://www.ti.com/lit/ds/symlink/cc1101.pdf.

[Texb]   Texas Instruments. CC2420 datasheet. http://www.ti.com/lit/ds/symlink/cc2420.pdf.

[Texc]   Texas Instruments. CC2520 datasheet. http://www.ti.com/lit/ds/symlink/cc2520.pdf.

[Texd]   Texas Instruments. CC430 datasheet. http://www.ti.com/lit/ds/symlink/cc430f5137.pdf.

[Texe]   Texas Instruments. MSP430F1611 datasheet. http://www.ti.com/lit/ds/symlink/msp430f1611.pdf.

[TPS+05]   G. Tolle, J. Polastre, R. Szewczyk, D. Culler, N. Turner, K. Tu, S. Burgess, T. Dawson, P. Buonadonna, D. Gay, and W. Hong. A macroscope in the redwoods. In *Proceedings of the 3rd ACM Conference on Embedded Networked Sensor Systems (SenSys)*, 2005.

[VVV07]   Z. Vincze, R. Vida, and A. Vidács. Deploying multiple sinks in multi-hop wireless sensor networks. In *Proceedings of the IEEE International Conference on Pervasive Services (ICPS)*, 2007.

[WALJ+06]   G. Werner-Allen, K. Lorincz, J. Johnson, J. Lees, and M. Welsh. Fidelity and yield in a volcano monitoring sensor network. In *Proceedings of the 7th USENIX Symposium on Operating Systems Design and Implementation (OSDI)*, 2006.

[WASW05]   G. Werner-Allen, P. Swieskowski, and M. Welsh. MoteLab: A wireless sensor network testbed. In *Proceedings of the 4th ACM/IEEE*

International Symposium on Information Processing in Sensor Networks (IPSN), 2005.

[WC02]   B. Williams and T. Camp. Comparison of broadcasting techniques for mobile ad hoc networks. In *Proceedings of the 3rd ACM International Symposium on Mobile Ad Hoc Networking and Computing (MobiHoc)*, 2002.

[WCB01]   M. Welsh, D. Culler, and E. Brewer. SEDA: An architecture for well-conditioned, scalable internet services. In *Proceedings of the 18th ACM Symposium on Operating Systems Principles (SOSP)*, 2001.

[WCL+07]   M. Wachs, J. I. Choi, J. W. Lee, K. Srinivasan, Z. Chen, M. Jain, and P. Levis. Visibility: A new metric for protocol design. In *Proceedings of the 5th ACM Conference on Embedded Networked Sensor Systems (SenSys)*, 2007.

[WHC+13]   Y. Wang, Y. He, D. Cheng, Y. Liu, and X.-y. Li. Triggercast: Enabling wireless constructive collisions. In *Proceedings of the 32nd IEEE International Conference on Computer Communications (INFOCOM)*, 2013.

[YI85]   S. Yoshida and F. Ikegami. A comparison of multipath distortion characteristics among digital modulation techniques. *IEEE Transactions on Vehicular Technology*, 34(3), 1985.

[YZLZ05]   F. Ye, G. Zhong, S. Lu, and L. Zhang. GRAdient broadcast: A robust data delivery protocol for large scale sensor networks. *ACM Wireless Networks (WINET)*, 11(3), 2005.

[ZF06]   Y. Zhang and M. Fromherz. Constrained flooding: A robust and efficient routing framework for wireless sensor networks. In *Proceedings of the 20th IEEE International Conference on Advanced Information Networking and Applications (AINA)*, 2006.

[ZFL+13]   M. Zimmerling, F. Ferrari, R. Lim, O. Saukh, F. Sutton, R. Da Forno, R. S. Schmidt, and M. A. Wyss. Poster abstract· A reliable wireless nurse call system: Overview and pilot results from a summer camp for teenagers with Duchenne muscular dystrophy. In *Proceedings of the 11th ACM Conference on Embedded Networked Sensor Systems (SenSys)*, 2013. To appear.

[ZFM+12]   M. Zimmerling, F. Ferrari, L. Mottola, T. Voigt, and L. Thiele. pTunes: Runtime parameter adaptation for low-power MAC protocols. In *Proceedings of the 11th ACM/IEEE International Conference on Information Processing in Sensor Networks (IPSN)*, 2012.

[ZFMT13]   M. Zimmerling, F. Ferrari, L. Mottola, and L. Thiele. On modeling low-power wireless protocols based on synchronous packet transmissions. In *Proceedings of the 21st IEEE International Symposium on Modeling, Analysis and Simulation of Computer and Telecommunication Systems (MASCOTS)*, 2013.

[ZG03]   J. Zhao and R. Govindan.   Understanding packet delivery performance in dense wireless sensor networks. In *Proceedings of the 1st ACM Conference on Embedded Networked Sensor Systems (SenSys)*, 2003.

[ZZHZ10]   T. Zhu, Z. Zhong, T. He, and Z.-L. Zhang. Exploring link correlation for efficient flooding in wireless sensor networks. In *Proceedings of the 7th USENIX Symposium on Networked Systems Design and Implementation (NSDI)*, 2010.

List of Publications

The following list includes publications that form the basis of this thesis.
The corresponding chapters are indicated in parentheses.

F. Ferrari, M. Zimmerling, L. Thiele, and O. Saukh. **Efficient network flooding and time synchronization with Glossy.** In *Proceedings of the 10th ACM/IEEE International Conference on Information Processing in Sensor Networks (IPSN)*. Chicago, IL, USA, April 2011. ***Best paper award.*** (Chapter 2)

F. Ferrari, M. Zimmerling, L. Thiele, and L. Mottola. **The bus goes wireless: Routing-free data collection with QoS guarantees in sensor networks.** In *Proceedings of the 4th International Workshop on Information Quality and Quality of Service for Pervasive Computing (IQ2S, in conjunction with IEEE PerCom)*. Lugano, Switzerland, March 2012. (Chapter 3)

F. Ferrari, M. Zimmerling, L. Thiele, and L. Mottola. **Poster abstract: The low-power wireless bus: Simplicity is (again) the soul of efficiency.** In *Proceedings of the 11th ACM/IEEE International Conference on Information Processing in Sensor Networks (IPSN)*. Beijing, China, April 2012. (Chapter 3)

F. Ferrari, M. Zimmerling, L. Mottola, and L. Thiele. **Low-power wireless bus.** In *Proceedings of the 10th ACM Conference on Embedded Networked Sensor Systems (SenSys)*. Toronto, Canada, November 2012. (Chapter 3)

F. Ferrari, M. Zimmerling, L. Mottola, and L. Thiele. **Virtual synchrony guarantees for cyber-physical systems.** In *Proceedings of the 32nd IEEE International Symposium on Reliable Distributed Systems (SRDS)*. Braga, Portugal, October 2013. (Chapter 4)

The following list includes publications that are not part of this thesis.

F. Ferrari, A. Meier, and L. Thiele. **Accurate clock models for simulating wireless sensor networks.** In *Proceedings of the 3rd International Workshop on OMNeT++ (OMNeT++ Workshop, in conjunction with SIMUTools).* Malaga, Spain, March 2010.

F. Ferrari, A. Meier, and L. Thiele. **Secondis: An adaptive dissemination protocol for synchronizing wireless sensor networks.** In *Proceedings of the 7th IEEE Conference on Sensor Mesh and Ad Hoc Communications and Networks (SECON).* Boston, MA, USA, June 2010.

M. Zimmerling, F. Ferrari, M. Woehrle, and L. Thiele. **Poster abstract: Exploiting protocol models for generating feasible communication stack configurations.** In *Proceedings of the 9th ACM/IEEE International Conference on Information Processing in Sensor Networks (IPSN).* Stockholm, Sweden, April 2010.

J. Beutel, B. Buchli, F. Ferrari, M. Keller, L. Thiele, and M. Zimmerling. **X-Sense: Sensing in extreme environments.** In *Proceedings of the Conference on Design, Automation and Test in Europe (DATE).* Grenoble, France, March 2011. Invited paper.

M. Zimmerling, F. Ferrari, L. Mottola, T. Voigt, and L. Thiele. **pTunes: Runtime parameter adaptation for low-power MAC protocols.** In *Proceedings of the 11th ACM/IEEE International Conference on Information Processing in Sensor Networks (IPSN).* Beijing, China, April 2012. *Best paper runner-up.*

O. Landsiedel, F. Ferrari, and M. Zimmerling. **Poster abstract: Capture effect-based communication primitives.** In *Proceedings of the 10th ACM Conference on Embedded Networked Sensor Systems (SenSys).* Toronto, Canada, November 2012. *Best poster award.*

R. Lim, C. Walser, F. Ferrari, M. Zimmerling, and J. Beutel. **Demo abstract: Distributed and synchronized measurements with FlockLab.** In *Proceedings of the 10th ACM Conference on Embedded Networked Sensor Systems (SenSys).* Toronto, Canada, November 2012.

R. Lim, F. Ferrari, M. Zimmerling, C. Walser, P. Sommer, and J. Beutel. **FlockLab: A testbed for distributed, synchronized tracing and profiling of wireless embedded systems.** In *Proceedings of the 12th ACM/IEEE International Conference on Information Processing in Sensor Networks (IPSN)*. Philadelphia, PA, USA, April 2013.

M. Zimmerling, F. Ferrari, L. Mottola, and L. Thiele. **On modeling low-power wireless protocols based on synchronous packet transmissions.** In *Proceedings of the 21st IEEE International Symposium on Modeling, Analysis and Simulation of Computer and Telecommunication Systems (MASCOTS)*. San Francisco, CA, USA, August 2013.

O. Landsiedel, F. Ferrari, and M. Zimmerling. **Chaos: Versatile and efficient all-to-all data sharing and in-network processing at scale.** In *Proceedings of the 11th ACM Conference on Embedded Networked Sensor Systems (SenSys)*. Rome, Italy, November 2013. *Best paper award.*

M. Zimmerling, F. Ferrari, L. Mottola, and L. Thiele. **Poster abstract: Synchronous packet transmissions enable simple yet accurate protocol modeling.** In *Proceedings of the 11th ACM Conference on Embedded Networked Sensor Systems (SenSys)*. Rome, Italy, November 2013.

M. Zimmerling, F. Ferrari, R. Lim, O. Saukh, F. Sutton, R. Da Forno, R. S. Schmidt, and M. A. Wyss. **Poster abstract: A reliable wireless nurse call system: Overview and pilot results from a summer camp for teenagers with Duchenne muscular dystrophy.** In *Proceedings of the 11th ACM Conference on Embedded Networked Sensor Systems (SenSys)*. Rome, Italy, November 2013.

Curriculum Vitæ

Personal Data

Name	Federico Ferrari
Date of Birth	January 6, 1981
Citizenship	Italian

Education

2008–2013	ETH Zurich Computer Engineering and Networks Laboratory Ph.D. under the supervision of Prof. Dr. Lothar Thiele
2006–2008	Università della Svizzera Italiana Advanced Learning and Research Institute (ALaRI) M.Sc. in Embedded Systems Design
1999–2003	Università di Bologna B.Sc. in Electrical Engineering
1994–1999	Liceo Scientifico Leonardo da Vinci, Cerea Secondary School with focus on Mathematics and Sciences

Professional Experience

2008–2013	ETH Zurich Computer Engineering and Networks Laboratory Research and teaching assistant
2005–2006	Università di Bologna Microelectronics Research Group Student intern

Honors and Awards

Nov. 2013	Best Paper Award at ACM SenSys 2013
Nov. 2012	Best Poster Award at ACM SenSys 2012
Apr. 2012	Best Paper Runner-up at ACM/IEEE IPSN 2012
Apr. 2011	Best Paper Award at ACM/IEEE IPSN 2011

www.ingramcontent.com/pod-product-compliance
Lightning Source LLC
Chambersburg PA
CBHW051505170526
45166CB00001B/392